TRANSACTIONS

OF THE

AMERICAN PHILOSOPHICAL SOCIETY

HELD AT PHILADELPHIA
FOR PROMOTING USEFUL KNOWLEDGE

NEW SERIES—VOLUME 38, PART 3

THE FREEZING OF SUPERCOOLED WATER

N. ERNEST DORSEY

National Bureau of Standards

THE AMERICAN PHILOSOPHICAL SOCIETY
INDEPENDENCE SQUARE
PHILADELPHIA 6

November, 1948

THE FREEZING OF SUPERCOOLED WATER

N. Ernest Dorsey

CONTENTS

A. INTRODUCTION

I. HOW THE WORK ORIGINATED

Something over a decade ago while I was engaged in the assembling and compiling of published data pertaining to the properties of water in its several phases and to its transition from phase to phase [29],[1] my interest in the supercooling and freezing of water was aroused. (A substance is said to be supercooled when it has been cooled below its melting point and is still liquid. Thus unfrozen water at temperatures below 0°C (32°F) is supercooled water. It has long been known that water can be greatly supercooled, − 20°C (− 4°F) and lower temperatures having been reached.)

My curiosity was first excited by the fact that apparently reliable observations reported many years before were entirely incompatible with certain current opinions and statements. For example, L. Dufour [30] had, among other things, reported that globules of water suspended in a liquid of the same density could be far supercooled although subjected to such violent disturbances as are produced by dropping various solid particles through them. He occasionally carried the supercooling to − 20°C. And it had been stated [60] that Pictet had found that a stoppered flask half full of water and cooled to − 19°C could be violently shaken without causing the water to freeze. And Despretz [26, 27] had studied the dilatation of supercooled water to − 20°C. In contrast to these, Barnes [9] had more recently written that agitation, particularly if the water contains dissolved air, makes supercooling almost impossible, that by avoiding agitation he had cooled water to − 6°C, but that below that temperature "the degree of instability reaches such a critical state . . . that extraordinary precautions have to be taken for further cooling." Others expressed similar views—that supercooled water is essentially unstable, that it is difficult to supercool water by more than a few degrees, and that quiescence is necessary.

Also, as a conclusion from some unsatisfactorily reported observations of his own, Barnes [12, 13] had stated that by a suitable procedure it is possible to deprive supercooled water of its trihydrol, and that at temperatures below 0°C a very appreciable interval—apparently an hour or more—is required for restoring the equilibrium. That such a long time should be required for the establishment of dissociative equilibrium caused surprise.

All these things—incompatibilities, the supposed evil effect of agitation, the supposed essential instability of supercooled water, the supposed excessive slowness with which dissociative equilibrium is established—and others, such as the importance very generally attached to the time required for freezing, the very slight attention given to the temperature to which a specimen can be supercooled, the widespread belief that only very

[1] Numbers in brackets indicate References at end of article.

small volumes of a melt can be much supercooled—all these things aroused my curiosity and raised many questions. They invited an independent survey of the subject, to be made, not with the idea of obtaining data that may be interpreted by an accepted theory, but with the idea of seeing for one's self how water actually does behave when cooled under various conditions, and how that behavior depends upon the source and the treatment of the water. And that invitation was strengthened by the fact that a crude preliminary test showed that water contained in a sealed glass bulb, which it half filled, could readily be supercooled to − 8°C, even when agitated.

Such was the origin of these studies, but it was nearly three years later before they could be undertaken.

II. GROWTH OF THE INVESTIGATION

The work herein reported has been done at such odd times as were available from other and more immediately important work. It seriously began in the summer of 1936, and observations were completely discontinued during five periods ranging in length from six months to two and one-half years, and for the sealed bulbs to be presently mentioned there were intermissions of nearly four years. The obvious disadvantages of such interruptions are in part offset by certain real advantages, but there is a dissipation of effort upon side issues, and new issues tend to crowd out older ones. The work being purely exploratory, plans could be made only for the immediate future. Also, the time that could be devoted to the work was very uncertain. Hence it seemed wise to avoid the construction of elaborate equipment, and to be content with simple, easily constructed apparatus, and with readily available makeshifts. That has been done.

At first, the primary purpose was to determine how variable might be the temperature to which a given specimen hermetically sealed in a glass bulb can be supercooled, and whether, and to what extent, the supercooling is affected by (1) agitation, (2) thermal shock, and (3) length of time that the specimen had been supercooled.

It was found, in general, that the extent of the supercooling of a given specimen of distilled water at a given time is fixed within moderately narrow limits, varies from specimen to specimen, and is essentially independent of the effects mentioned. As the specimen is slowly cooled, nothing appears to happen until a certain temperature is reached, and then, in general, almost in the twinkling of an eye, the entire volume spontaneously solidifies. That temperature will be called the spontaneous-freezing-point, or the temperature of spontaneous freezing, and will be denoted by t_{sf}. Of course only a small fraction of the initial water is actually ice at the conclusion of this rapid freezing, but the ice crystals are so numerous and intermingled and are distributed so uniformly throughout the entire volume that the whole is locked together into a single solid mass.

Then arose the questions: What is the t_{sf} of natural waters? Is it ever as high as 0°C (32°F) and if so, under what conditions?

So specimens of natural waters from many sources were studied. Some contained organic matter; in some, organic growth was discouraged by the presence of metallic copper. In no case did a specimen freeze spontaneously above − 3°C.

Various specimens of distilled and of redistilled water, both air-saturated and air-free, were studied, some of which were repeatedly supercooled to or below − 20°C. The presence of air produced no significant effect; the effect, if any, of a prolonged preheating was not annulled by subsequent freezings.

The observations indicated, in general, that the so-called nuclei or germs involved in the initiation of freezing do not arise directly from the substance of the water, but probably are, or arise from, foreign motes, either suspended in the water or attached to its boundaries. Hence experiments were made to see whether nominally clean water might not be so heterogeneous in mote content that two samples, even of several cubic centimeters each, drawn from the same volume would freeze spontaneously at different temperatures. Such heterogeneity was found, and the spontaneous freezing of mixtures of waters was studied.

A pouring of the water from end to end of the bulb, a tapping of the outside of the bulb, the impact of solids immersed in the supercooled water, were found, within the range studied, to be without effect upon the initiation of freezing. But a very gentle rubbing together of two solids immersed in supercooled water, and a certain type of violent splashing, may initiate freezing at a temperature above that at which spontaneous freezing occurs.

Such were the results obtained during the first period of study, and given in a preliminary report [28] published in 1938. During the next year the work was entirely suspended except for a prolonged chilling of certain of the bulbs.

Then a number of experiments were made to determine whether, and to what extent, the placing of various kinds of solids in water affects the temperature at which it freezes spontaneously. The results were erratic; as one would expect if motes initiate spontaneous freezing.

Practical questions arose as to whether the strength with which ice adheres to a substance depends upon the nature of the substance, and if so, whether it is associated with the efficacy of the substance in the initiation of freezing by rubbing the substance against itself while immersed in supercooled water. And this led to an extended study of the initiation of freezing by rubbing, and of the effect upon it of solutes. A question regarding the possibility of squirting supercooled water through a nozzle arose, and was studied, as was the

effect of solutes upon the temperature at which water freezes spontaneously.

Prior to 1942 the work was done in a fourth floor, steam-heated room with linoleum covered floor. At the end of 1941 a change was necessary in order to accommodate war work, and this work was moved to a first-floor, air-heated room with uncovered and unpainted concrete floor. The air of that room was found to be exceedingly dusty. In what follows, that will be called the dusty room.

During the entire course of the work, the sealed specimens were refrozen from time to time, and many incidental observations were made; also various side issues bearing on the work were studied as they arose.

On later pages, the data obtained, especially those relating to the sealed bulbs, are presented very fully, in order that the reader may see how the several specimens actually behaved, and may be able to check the data against such ideas and theories as may occur to him.

Only near the conclusion of the observations were the results reported by previous workers on other melts and on solutions studied, were the results herein reported carefully compared with those demanded by current theories, and was a more satisfactory theory, which had been gradually taking shape, worked up.

Those interested in the growth and solution of crystals are referred to Spangenberg's summary [75] of such work prior to 1935.

B. EXPERIMENTAL DATA

I. THERMAL INITIATION OF FREEZING

1. TEMPERATURE OF SPONTANEOUS FREEZING

a. Bulbs and Their Cleaning

(1) *Bulbs.*—At the very beginning of the work (August 26, 1936) it was decided to seal the specimens of water in glass bulbs so as to avoid every possibility of a progressive change arising from air-borne materials. Glass was chosen because it was readily obtainable, easily worked, and afforded a clear view of the contents. Questions regarding the possible effect of the glass upon the specimen naturally arise, and must be considered, but they can with profit be deferred until after one has learned how water in glass behaves. It may, however, be remarked here that nothing has been observed that indicates that the glass itself affects the phenomena under study. At the beginning, soft soda glass was used; later, Pyrex was also used, and certain experiments with quartz bulbs closed by a ground-in stopper were made.

The several types of bulb are shown in figure 1. The *sealed bulbs,* used from the beginning for specimens that were to be studied intensively over long periods, are shown at *A*. Each bulb is cylindrical and has hemispherical ends, from one of which extends axially a 7-mm tube through which the specimen of water is introduced. The tube is then sealed by fusion at a point a centimeter or two above the bulb. The bulbs proper are externally about 2 cm in diameter by 3.5 to 4.5 cm long, their internal volumes being about 6 to 8 milliliters.[1] They were usually about half filled.

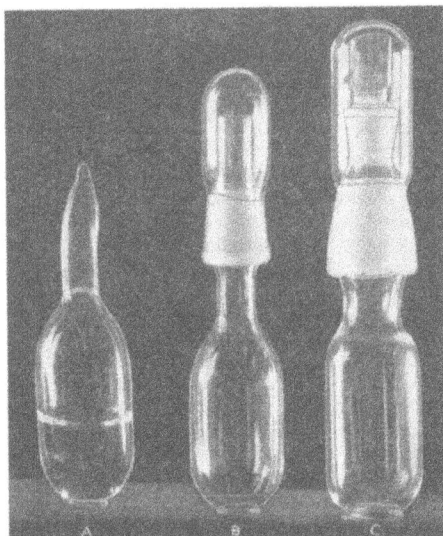

FIG. 1. Types of bulbs. *A*, sealed; *B*, capped; *C*, stoppered and capped.

The soft glass bulbs are serially numbered, and both in the laboratory records and in this report are severally distinguished by a capital *C* prefixed to the appropriate number. There are two sets of sealed bulbs of Pyrex. Those of one are marked with a capital *P* followed by an arabic numeral, and are so designated in this report; those of the other are marked with green arabic numerals, and are herein designated by a capital *G* prefixed to the numerals.

Eighty-four hermetically sealed specimens have been intensively studied. Their contents and histories are given in the appendix, and their spontaneous-freezing-points are shown graphically in figures 3 to 27, where every freezing after December 19, 1936, is accounted for. Several other bulbs were prepared for special purposes and were studied less intensively. Information regarding them and the data so obtained will be found in appropriate places.

The bulbs shown at *B* in figure 1 are herein called *capped bulbs,* and are severally designated by a capital letter (*A* to *I*) scratched on the bulb. They are all of Pyrex. The long neck extending into the cap enables

[1] Not 16 to 17, as stated in the 1938 paper.

one to detect quickly any leakage from the chilling bath through the ground joint. Only very seldom was there a perceptible leakage.

The *stoppered and capped bulbs* are shown at *C* in figure 1. There are two sets of six bulbs each. One set is of Pyrex, the bulbs being serially marked with arabic numerals; the other is of quartz and the bulbs are marked with roman numerals. In the laboratory records and in this report they are designated by an *SC* (stoppered and capped) prefixed to the proper numeral.

The quartz bulbs are a little larger than the others, all of which are of approximately the same size as the sealed bulbs.

(2) *Cleaning the Bulbs.*—Throughout the work, the bulbs were carefully cleaned before being charged with the specimen. They were washed, then soaked, usually over night, in a strong chromic cleaning solution (potassium bichromate and concentrated sulphuric acid, with such water as may have been acquired from the air or from moist bulbs previously placed in it), then washed with distilled water, and inverted for fifteen minutes or so over a jet of live steam. The steaming was always continued until the walls seemed to wet uniformly. If that condition was not secured within fifteen minutes, the bulb was again cleaned with the chromic solution, washed, and steamed again. This was continued as often as necessary in order to secure uniform wetting. Generally, one cleaning in the chromic solution was sufficient; very seldom were more than two required. When the bulb was removed from the steam jet it was at once either charged and sealed, or otherwise closed, or was placed inverted in a closed cabinet. In the latter case, it was usually steamed again just before being charged.

At the beginning of the work, before the part played by motes was clearly understood, it was thought that such cleaning would be entirely satisfactory for every purpose; and that seemed to be borne out by the observations. But it is obvious that as soon as the steaming of the bulb is stopped, air rushes in, carrying its full complement of motes, some of which will adhere to the moist wall. When a specimen is placed in such a bulb it will inevitably take up some of the motes in the contained air, and its properties will also be affected by the motes that are adhering to the wall. Hence it will be impossible by such procedures to determine the properties of specimens that are totally free from motes of the kind present in the air in which the bulb was steamed.

Were one trying to determine to how low a temperature mote-free water can be supercooled, he would have to take much greater precautions than those taken in this work.

However, in this study one is not concerned with the question: To how low a temperature can water be supercooled? but with the more practical ones: What determines the temperature at which a given specimen of water freezes spontaneously, and how can that temperature be deliberately changed? One is working with mote-infested waters, not with mote-free water. The problem is to see how the temperature at which such infested water freezes spontaneously varies from specimen to specimen, how that temperature can be varied,

TABLE 1

EFFECTIVENESS OF CLEANING

Pairs of capped bulbs—A and C, or C and D—were treated in the same way and then charged with about 4 ml of water taken from the same source, and the spontaneous-freezing-points of their contents were determined. If the water were strictly homogeneous, if the bulbs were thoroughly clean, and if the specimens took up no motes as they were being placed in the bulbs, then the two specimens would freeze spontaneously at the same temperature. Differences in those temperatures may be taken as an upper limit to those arising from differences in the cleaning of the bulbs. The entries "washed" and "rinsed" indicate that, just before charging, the bulb was washed or rinsed with water from the source used in charging it. The greatest difference is 4°C (water d) after a mere rinsing; the next is 3.5°C (water g) after a rinsing; then 3.1°C and 3.2°C (water g) after cleaning without steaming. After steaming the difference does not exceed 0.6°C.

Temperatures in °C

Bulb	A	C	A	C	A	C
Treatment	Spontaneous-Freezing-Point					
....	Water (a)		—	—	—	—
Washed.......	−13.4	−12.0*	Water (b)		—	—
Cleaned and steamed.....	—	—	−12.8	−12.5	Water (c)	
Washed.......	—	—	—	—	−13.2	−14.6
Rinsed.......	—	—	—	—	−13.0	−14.0
Rinsed.......	—	—	—	—	−13.2	−14.7*
Washed and steamed.....	Water (d)		—	—	−14.1	−13.7
Steamed......	−13.0	−13.6*	Water (e)		—	—
Rinsed.......	—	—	−12.8	−12.9	—	—
Rinsed.......	—	—	−14.3	−14.4	—	—
Rinsed.......	—	—	−14.0	−14.2	—	—
Rinsed.......	—	—	−14.0	−14.0	—	—
Rinsed.......	−14.2	−10.2	—	—	—	—
Rinsed.......	−14.8	−14.4*	—	—	Water (f)	
Rinsed.......	Water (g)		—	—	− 6.9	− 6.9
Rinsed.......	−12.9	−11.2	—	—	—	—
Rinsed.......	−12.2	−10.9	—	—	—	—
Rinsed.......	−13.3	− 9.8	—	—	—	—
Cleaned......	−13.6	−15.0	—	—	—	—
Rinsed.......	−15.0	−14.5	—	—	—	—
"f" replaces "g"	—	—	—	—	− 6.8	− 6.8
Cleaned......	−12.5	−14.0	—	—	—	—
Rinsed.......	−13.3	−15.0	Water (h)		—	—
Rinsed.......	—	—	− 6.6	− 7.0	—	—
Cleaned......	−14.8	−15.0	—	—	—	—
Rinsed.......	—	—	− 7.1	− 6.9	—	—
Cleaned......	−15.0	−14.0*	—	—	—	—
Cleaned......	−15.8	−12.7	—	—	—	—
Repetition.....	−15.5	—	—	—	—	—
Rinsed......	—	−15.0*	—	—	—	—
Cleaned......	−12.0	−15.2	—	—	—	—
Cleaned......	−15.4	—	—	—	—	—
Rinsed.......	—	—	− 7.0	− 7.0	—	—
Cleaned......	—	—	− 7.0	− 9.2	—	—

Temperatures in °C—*Continued*

Bulb	C	D	C	D	C	D
Treatment	Spontaneous-Freezing-Point					
....	Water (g)		—	—	—	—
Cleaned......	−15.6	−15.0*	—	—	—	—
Cleaned......	−13.5	−13.2	Water (i)		—	—
Cleaned......	—	—	−14.0	−14.7*	—	—
Cleaned......	—	—	−14.0	−13.6	—	—
Rinsed.......	—	—	−14.9	−14.9	Water (j)	
Rinsed.......	—	—	—	—	− 6.8	− 7.0
Cleaned......	Water (k)		−14.4	−14.3*	—	—
Rinsed.......	− 6.4	− 6.0	—	—	—	—
Cleaned......	—	—	−14.5	−14.8	—	—

* Other waters were frozen in the bulbs between this entry and that on the next line.

what effects are produced by treatments of various kinds, whether freezing can be initiated by mechanical means, and if so, how, etc.; and to explain the whole in some logical manner. Hence, if the motes on the bulb and in its contained air produce effects that are small as compared with the corresponding differences between the several specimens of water, their presence does no more than modify the specimen, in effect, replacing it by a slightly different one. A change that is of no consequence.

Even the earliest determinations of the temperatures of spontaneous freezing of specimens of water enclosed in sealed bulbs indicated that differences that could be attributed to the bulbs alone must be very small as compared with many of those attributable to the specimens themselves. From later observations, taken for another purpose, one can form an estimate of the magnitude of the effect that may be produced by the motes introduced by the bulb. Typical data are given in table 1. It will be noticed that, after steaming, the spontaneous-freezing-points of the specimens in the bulbs being compared do not differ by more than 0.6°C; after cleaning without steaming, the greatest difference is about 3°C, 7 out of 13 ranging from 0.1 to 0.7°C and averaging 0.4°C; and with one exception, a second rinsing with the water from which the specimen was taken sufficed to reduce the difference to 0.5°C or less. Eleven distinct lots of water, freezing spontaneously at temperatures ranging from − 6 to − 15°C, were used in obtaining the data given in the table. Each tabulated temperature corresponds to a different specimen of water; the temperatures are averages, usually of 3 or 4, and are arranged vertically in chronological order. These observations were made in the steam-heated, linoleum covered floor, fourth-story room occupied prior to 1942.

Whence it seems that bulbs cleaned prior to 1942 may in general be expected to yield comparable data, differences as great as 3°C being very exceptional. That is sufficient for the present study.

Attempts to obtain similar data in the air-heated, dusty, first-story room, with unpainted and uncovered concrete floor, occupied after 1941, failed. Only rarely did two bulbs cleaned and steamed, and charged with the same water, give the same result. And there were many indications that extraneous motes were usually introduced whenever the bulbs were opened, even though such opening was done inside an assumed dust-proof box lined with black slipper rayon and provided with rayon sleeves. A washing of the floor just before work gave little or no improvement; and the results in summer were not significantly better than those in winter.

It seems that motes caught by the wet walls of a bulb may, after drying, adhere so strongly as to resist washing and even a hurried cleaning. Possibly the surface tension of the evaporating film of water between them and the glass draws them down almost into molecular contact with the glass, the union being something of the nature of that between two optical flats that have been wrung together. And surely they become cemented to the glass by the deposition of those very slightly soluble substances that are dissolved in the evaporating water. The removal of that very insoluble cement is slow and difficult. As will be seen presently, the solution of motes suspended in the water may go on for years.

After standing at rest for a time, some of the charged and sealed bulbs, especially those of Pyrex, ceased to be uniformly wettable on their inner surfaces above the water-line. Of course, the walls of the charged bulbs of soft glass became in time etched, producing flocculent material in the water.

b. Procedures and Preliminary Results

The prepared bulbs were kept at laboratory temperatures except when undergoing thermal treatment—heating, prolonged cooling, or freezing.

When a specimen was to be frozen, its bulb was generally suspended by a suitably weighted harness in alcohol, which was cooled as desired. Prior to December 19, 1936, the alcohol was contained in a large test tube closed by a paraffined cork, and inserted into a closely fitting cylinder of metal or gauze, which was itself in an ordinary Thermos bottle containing in the earlier experiments a mixture of crushed ice, salt, and water, and in the later ones alcohol, cooled by suitable additions of solid CO_2. There was a thermometer in the chilling mixture, and another in the alcohol in which the bulb was suspended. It was assumed that the temperature of the water in the bulb was the same as that of the thermometer in the alcohol with it.

Although quite unsatisfactory, in that neither the thermometer nor the specimen could be continuously watched, and in that the adjustment of the temperature of the chilling mixture was haphazard, this arrangement sufficed to show that, contrary to the common opinion, primary significance does not attach to the time required for a specimen of water to freeze, but to the temperature attained by the specimen, and that quiescence is not essential to supercooling. For example, in the very first experiment six hours were taken to cool a certain bulb (CI) from 23°C to -9.5°C; it was then held over night at -9.5 to -8°C, and rose to -6°C while the ice and salt mixture was being replaced, and then in 40 minutes it was cooled from -6°C to -14.5°C, at which temperature the water froze. It was then melted, cooled rapidly to -13.5°C, kept at -14 to -14.2°C for 6 minutes, and then cooled slowly for 2 minutes to -14.7°C, at which temperature it froze. The two temperatures of freezing are essentially the same, although in one case 24 hours were taken for the cooling, the water having been continuously between -6 and -9°C for 18 hours, whereas in the second case the total duration of chilling was only 10 minutes. In the first case, after the bulb had remained over night at -9.5 to -8°C, it was violently shaken, but did not freeze. The next night the specimen was held between -13 and -10°C for 17 hours without freezing.

Many similar experiments yielding similar results were made, sufficient to convince me that, in spite of many observed irregularities, each specimen, so long as it is unchanged, freezes spontaneously at a definite temperature, fixed within a fraction of a degree, and essentially unaffected by the time taken by the specimen to reach that temperature. Furthermore, the specimen's size and quiescence, or the lack of it, have little if any effect upon its t_{sf}, its temperature of spontaneous freezing, and different specimens may differ widely in the respective temperatures at which they freeze spontaneously.

Thus was opened an extended vista inviting exploration, and improved facilities seemed justified. The Thermos bottle was replaced on December 19, 1936, by an unsilvered Dewar cylinder, for which I am indebted to the late Mr. E. O. Sperling, glass blower for the Bureau. Later, unsilvered "laboratory exhausted" Thermos fillers of various sizes were also used. The specimen could now be kept under observation continuously, and the temperature of the alcohol at the instant of freezing could be observed. Also, the specimen could be observed at any time with a low-power microscope, and could be examined between crossed Polaroids. Each was done, without adding significantly to what could be determined by the unaided eye. Such unsilvered containers were used throughout the rest of the work. The results so obtained are so far superior to those obtained with the Thermos bottles, which however they confirm, that nothing more need be said of the earlier work.

The alcohol contained in the unsilvered Dewar cylinder (which term includes the Thermos fillers) will be called the chilling bath, or simply the bath. It was always cooled by frequent additions of small pieces of solid CO_2, and was stirred by dry air bubbling from an aperture in the end of a glass tube reaching to, or nearly to, the bottom of the cylinder, or occasionally by a manually operated mechanical stirrer. In the bath hung a thermometer with its bulb at approximately the same level as the center of the specimen bulb, and not far from it. The specimen bulb was usually suspended directly in the chilling bath, but occasionally it was suspended in alcohol contained in a large test tube, which was itself suspended in the bath. In that case, a second thermometer was placed in the test tube, with its bulb adjacent to that containing the specimen of water. Results were the same in both cases.

c. Thermometers and Temperatures

In view of the nature of the work and the results, it seemed superfluous to have the thermometers carefully studied. They were all good thermometers, graduated to 1°C, distance between graduations being about 1.7 mm; and they were occasionally intercompared in the laboratory without discovering a discrepancy exceeding 0.1°C.

As the temperature corresponding to the specimen's t_{sf} was approached, the temperature of the bath was varied slowly, and it has been assumed throughout that the significant temperature of the specimen at the instant of freezing was that indicated by the thermometer adjacent to the bulb, which is the temperature here recorded as t_{sf}, the temperature of spontaneous freezing.

Actually, the temperature of the water was never uniform throughout, and at every point it was higher than that of the bath, except possibly in those few cases in which the temperature of the bath was slowly rising. However, we are not concerned with the exact temperature at which the specimen freezes, but merely with the variations in that temperature from time to time and from specimen to specimen. Hence, if as the freezing temperature is approached, specimens of the same size are cooled, always at approximately the same rate, it is improbable that the observed differences in the thermometer temperatures will differ significantly from the corresponding differences in the true values of t_{sf}. Indeed it has been found that if the specimen has been in the bath for at least 5 or 6 minutes during which time the temperature of the bath has varied not more than a degree or two, and if during the 2 or 3 minutes immediately preceding the freezing the temperature of the bath has varied not more than a few tenths of a degree per minute, then the thermometer temperature when freezing occurs is essentially the same as if the cooling had been much slower.

Nevertheless it seemed well to try to get some indication of the difference that might exist between the temperatures of the water and the bath. To that end the neck of one of the bulbs of the same lot as was used for the specimens was greatly extended, so that it

reached well above the surface of the alcohol when the bulb was immersed to the depth commonly used throughout this work. A charge of water, changed from time to time, was placed in the bulb and one terminal of a thermocouple [2] was immersed in it, the leads passing out through the long unsealed neck. The other terminal was tied to the bulb of the thermometer in the bath. Thus the excess of the temperature of the water above that of the bath could be determined. But it should be remembered that the presence of the leads will modify both the temperature and its distribution, which is perhaps the reason why no certain difference was observed when the position of the junction was changed.

A few typical observations so obtained are shown in figure 2, in which the circles represent the readings of

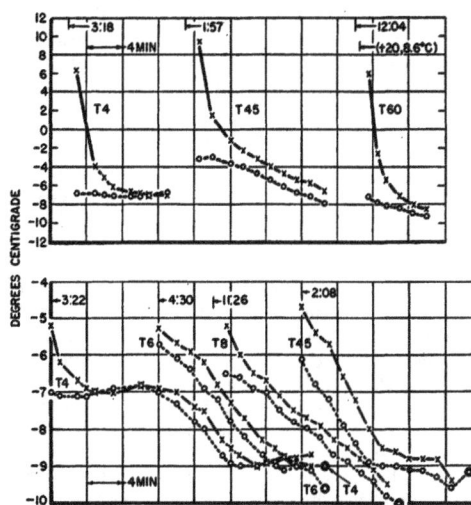

FIG. 2. Rates of cooling. Showing the rate at which the temperature of the water in the bulb (cross) approaches that of the thermometer in the bath (dot). A circle indicates that the specimen has frozen. For each pair of graphs, the clock time corresponding to a single abscissa is indicated.

the thermometer in the bath, and the crosses the corresponding temperatures of the junction in the water. When the terminal point is a double circle, it represents the thermometer reading immediately after the water froze; the corresponding water temperature could not be determined on account of the marked rise in temperature that accompanies freezing. The rate of cooling was intentionally varied widely.

The graphs in the upper section of the figure correspond to the first few observations after the bulb was

[2] For the thermocouple and its attendant apparatus, I am indebted to my colleague, Mr. E. F. Mueller and Mr. A. I. Dahl.

immersed in the alcohol. Those in the lower section correspond to the last, extending to the spontaneous freezing of the water; they are plotted on a much more open temperature scale. It will be noticed that even in the extreme case represented by $T60$ in which the bulb at 20°C was plunged into alcohol at $-8.6°C$ and the bath was being cooled at the rate of about 1°C in 4 minutes, a steady state corresponding to a temperature difference of about 1°C was reached within 5 minutes. When the temperature of the bath is kept constant within a few tenths of a degree, the water soon comes to essentially the same temperature as the bath.

These observations strengthen the opinion obtained from the long series of freezings; viz., that the spontaneous-freezing-point, as defined by the thermometer, is probably correct within a few tenths of a degree, and only in exceptional cases is it likely to be in error by so much as half a degree. In many cases the rate of cooling just before freezing occurred was so small and had been so for such a period that the error was probably not greater than a tenth of a degree.

d. Chance Enters

Although the temperature at which a given specimen was observed to freeze spontaneously was in general narrowly defined over a reasonable length of time, variations did occur, and for certain specimens those variations were quite marked.

In conformity with the fashion of the times, one might at once conclude that these variations are to be explained on statistical grounds, the initiation of freezing being a "chance" event. And such suggestion has been offered. But that solution is too easy. It should not be accepted except as a last resort. To do otherwise is to discourage investigation of the phenomenon.

That "chance" enters is obvious. It enters in several ways. If the initiation of freezing at a given temperature is conditioned by the presence of motes of a certain type or size, as was suggested by the work reported in 1938 and as is in conformity with the more recent work now to be reported, then every sample of water is distinctly heterogeneous and two specimens taken from the same sample will differ more or less, the difference being conditioned by the "chance" of the motes in one differing significantly from those in the other. Again, since observations must be made while the temperature is slowly falling, the temperature of the specimen will not be uniform throughout, and whether the specimen will freeze at a given instant will depend upon whether the most efficient mote is or is not in a region in which the temperature is at or below that at which that mote becomes effective. That is in part a matter of "chance." Still again, the initiation of freezing by a given mote at a given instant depends in part on the particular types and strengths of the molecular impacts to which it is subjected during a brief interval centered on that instant. And that is also a matter of "chance."

It is obvious that "chance" enters. It is not a question of demonstrating that it enters, but of determining the extent of the variations that it may impose upon the observations. The last is a proper subject for experimental study. Indeed it is only by experimental study that the extent of such variations can be determined prior to a far more complete understanding of the phenemenon than any we had at that time. Consequently, in what follows "chance" will seldom be mentioned. Attention will be concentrated upon (1) the reproducibility of the spontaneous-freezing-point for a given specimen, (2) the variation in that point from specimen to specimen, (3) its secular variation, and (4) means by which it may be intentionally changed.

e. Typical Observations and Conclusions

In table 2 are given examples of sets of observations as actually made. In the first column of each set is given the time of day at which the alcohol bath had the temperature given in the next column. In the third column are certain notes: "$C10$ put in," "Both in," etc. mean that immediately after reading the temperature recorded in the preceding column the bulb $C10$, or both bulbs, etc., was put into the cold alcohol; CO_2 means that solid CO_2 was added to the alcohol immediately after the temperature had been read, marks "do." have their usual significance, a dash means that no CO_2 was added; "Frozen," "$P24$ fr.," etc. means that the water in the bulb was frozen when the temperature was read, or froze immediately thereafter. At the beginning of each set are given the date of the observations, the designation of the specimen by means of the bulb number (see appendix for details), and the type of water; and at the end of each set are recorded the several observed values of the spontaneous-freezing-point, t_{sf}.

The set for $C10$, December 19, 1936, is the first in which an unsilvered Dewar cylinder was used. The technique is not as good as for the others; the pieces of CO_2 were too large, and they were not put into the alcohol promptly enough after the temperature was read. As a result, the CO_2 had not always completely vaporized before the next temperature was read. Consequently, there was at that time descending currents of cold alcohol, which may have caused the thermometer reading to be too low, or may have cooled a portion of the bulb significantly below the temperature indicated by the thermometer. This condition was corrected after a little experience.

In the set of October 1, 1943 two bulbs were in the cold bath at the same time; in the others, only one. In that set it should be noticed that at its first freezing (12:51) $G5$ had been continuously in the cold bath since 11:52, that the bath had never during that time been above $-10.7°C$ (11:53), and had on two previous occasions (12:32 and 12:47) been as low as $-16.7°C$ without the freezing of $G5$; nevertheless, $G5$ did freeze at -16.9 and at $-17.1°C$. Furthermore,

TABLE 2

TYPICAL SETS OF OBSERVATIONS

The temperatures recorded are those of the alcohol bath in which the sealed bulb is suspended. The bath is stirred by a stream of dry air; solid CO_2 in small pieces is added as indicated, a dash indicating that none is added. Temperature of spontaneous freezing $\equiv t_{sf}°C$.

December 19, 1936 C10 Conductivity			December 16, 1936 Conductivity—Continued		
Time	Temp.	Remarks	Time	Temp.	Remarks
...	−12.1	C10 put in	10:59	−11.6	do.
10:04	−11.1	—	11:00	−11.7	—
:05½	−11.0	CO_2	:01	−11.6	CO_2
:06	−11.1	—	:02	−11.7	—
:07	−11.0	CO_2	:03	−11.7	—
:08	−11.1	do.	:04	−11.6	CO_2
:09	−11.3	—	:05	−11.5	do.
:10	−11.2	CO_2	:06	−11.6	do.
:11	−11.2	do.	:07	−11.8	—
:12	−11.2	do.	:07½	−11.9	Frozen^a
:13	−11.5	Frozen^a	:12	−11.6	C10 put in
:17	−11.0	C10 put in	:13	−11.1	CO_2
:18	−10.0	CO_2	:14	−11.2	do.
:19	−9.9	do.	:15	−11.5	—
:20	−10.0	do.	:16	−11.5	CO_2
:21	−10.3	do.	:17	−11.6	do.
:22	−10.5	do.	:18	−11.8	—
:23	−10.7	—	:19	−11.8	CO_2
:24	−10.8	—	:20	−11.7	do.
:25	−10.7	CO_2	:21	−11.7	do.
:26	−10.7	do.	:22	−11.9	—
:27	−10.8	do.	:23	−11.8	CO_2
:28	−10.9	do.	:24	−11.8	Frozen^a
:29	−11.0	—	—	—	C10 put in
:30	−11.0	CO_2	:29	−10.1	CO_2
:31	−11.0	do.	:30	−10.1	do.
:32	−11.1	do.	:31	−10.7	do.
:33	−11.1	do.	:32	−11.0	do.
:34	−11.2	—	:33	−11.2	do.
:35	−11.1	CO_2	:34	−11.4	do.
:36	−11.1	do.	:35	−11.7	do.
:37	−11.1	do.	:36	−11.9	—
:38	−11.3	—	:37	−11.8	CO_2
:39	−11.4	—	:38	−11.7	do.
:40	−11.3	CO_2	:39	−11.7	do.
:41	−11.3	—	:40	−11.7	do.
:42	−11.4	—	:41	−11.7	do.
:43	−11.3	CO_2	:42	−12.0	—
:44	−11.2	do.	:43	−11.9	—
:45	−11.4	—	:44	−11.7	CO_2
:46	−11.3	CO_2	:45	−11.8	do.
:47	−11.3	do.	:46	−11.8	do.
:48	−11.3	do.	:47	−11.9	do.
:49	−11.3	do.	:48	−12.0	do.
:50	−11.6	—	:49	−12.0	do.
:51	−11.5	CO_2	:50	−12.1	Frozen^b
:52	−11.5	—			
:53	−11.5	CO_2		t_{sf}	
:54	−11.6	—		−11.5°C	
:55	−11.5	CO_2		−11.9	
:56	−11.6	—		−11.8	
:57	−11.5	CO_2		−12.1	
:58	−11.6	do.			

a Removed, melted, and returned to bath.
b Removed.

TABLE 2—Continued

October 1, 1943 P24, G5, P26 Distilled			October 1, 1943 Distilled—Continued		
Time	Temp.	Remarks	Time	Temp.	Remarks
11:52	−11.2	Both P24 and G5 in	1:00	−16.6	do.
:53	−10.7	CO_2	:01	−16.6	do.
:56	−13.2	do.	:02	−16.8	do.
:58	−13.6	do.	:03	−17.0−	do.
12:00	−13.8	do.	:04	−17.1	G5 fr.^b
:01	−14.1	P24 fr.^c			

	P24 t_{sf}	P26 t_{sf}	G5 t_{sf}
	−14.1	−16.7	−16.9
	−13.9	−16.7	−17.1

October 1, 1943 P24, G5, P26 Distilled (cont.)		
Time	Temp.	Remarks
:04	−13.6	P24 put in
:05	−13.0	CO_2
:06	−13.0	do.
:07	−13.2	do.
:08	−13.7	do.
:09	−13.7	do.
:10	−13.9	do.
:11	−13.9	P24 fr.^d
:13	−13.5	P26 put in
:14	−13.0	CO_2
:15	−13.0+	do.
:16	−13.2	do.
:17	−13.3	do.
:18	−13.6	do.
:19	−13.8	do.
:20	−14.0	do.
:21	−14.0	do.
:22	−14.2	do.
:23	−14.7	do.
:24	−14.8+	do.
:25	−15.0	do.
:26	−15.2	do.
:27	−15.4	do.
:28	−15.8	do.
:29	−15.9	do.
:30	−16.0+	do.
:31	−16.1	do.
:32	−16.4	do.
:32+	−16.7	P26 fr.^e
:37	−15.9	P26 put in
:38	−15.2	CO_2
:40	−15.4	do.
:41	−15.4	do.
:42	−15.7	do.
:43	−15.9	do.
:44	−16.1	do.
:45	−16.2	do.
:46	−16.7	do.
:47	−16.7	P26 fr.^f
:48	−16.3	CO_2
:49	−16.4	do.
:50	−16.6	do.
:51	−16.9	G5 fr.^g
:54	−16.3	G5 put in
:55	−15.9	CO_2
:56	−15.9	do.
:57	−16.0−	do.
:58	−16.1	do.
:59	−16.3	do.

January 19, 1937 C12 Distilled		
Time	Temp.	Remarks
10:11	+22.0	C12 put in
:21	−0.9	CO_2
:25	−7.8	do.
:26	−8.1	do.
:27	−8.9+	do.
:31	−10.0	do.
:32	−11.7	do.
:33	−13.5	do.
:34	−13.9	do.
:35	−13.9+	do.
:35½	−14.0	Frozen^a
:38	−13.7	C12 put in
:39	−13.0	CO_2
:40	−12.7	do.
:41	−13.0	do.
:42	−13.5	do.
:43	−13.6	do.
:44	−13.8	do.
:45	−13.8	do.
:46	−13.8	do.
:47	−13.8	do.
:48	−13.8	do.
:49	−13.8	do.
:50	−14.0	do.
:50½	−14.0	Frozen^b

t_{sf}
−14.0°C
−14.0

July 16, 1943 P25 Distilled		
Time	Temp.	Remarks
9:57	+10.0	P25 put in
:58	+10.3	CO_2
10:06	−1.4	do.
:09	−3.8	do.

c Removed P24, melted, returned to bath.
d Removed P24, replaced by P26.
e Removed P26, melted, returned to bath.
f Removed P26 and proceeded.
g Removed G5, melted, returned to bath.

TABLE 2—Continued

July 16, 1943 Distilled—Continued

Time	Temp.	Remarks
10:11	− 5.0	do.
:13	− 6.2	do.
:15	− 7.0	do.
:16	− 7.2	do.
:17	− 7.8	do.
:18	− 8.2	do.
:19	− 8.8	do.
:20	− 9.1	do.
:21	− 9.4	do.
:22	− 9.7	do.
:23	− 9.8+	do.
:24	− 9.9	Frozen[a]
:26	− 9.5	P25 put in
:27	− 9.0	CO2
:28	− 9.0	do.
:29	− 9.1	do.
:30	− 9.4	do.
:31	− 9.6	do.
:32	− 9.6+	do.
:33	− 9.7	do.
:34	− 9.8	do.
:35	−10.0−	do.
:36	−10.0+	do.
:37	−10.1	do.
:37½	−10.1+	Frozen[a]
:42	− 9.7	P25 put in
:43	− 8.6	CO2
:44	− 8.6	do.
:45	− 9.0	do.
:46	− 9.1	do.
:47	− 9.4	do.
:48	− 9.6	do.
:49	− 9.8	do.
:50	− 9.9	Frozen[b]

t_{sf}

− 9.9
−10.1
− 9.9

March 24, 1937 C42 Aquarium

Time	Temp.	Remarks
2:27	− 9.5	C42 put in
:28	− 8.3	CO2
:29	− 8.6	do.
:30	− 9.0+	do.
:31	− 9.9+	do.
:32	−10.4	—
:33	−10.3	CO2

March 24, 1937 Aquarium—Continued

Time	Temp.	Remarks
2:34	−10.7	Frozen[a]
:37	−10.2	C42 put in
:38	− 9.4	CO2
:39	−10.0	do.
:40	−10.3	—
:41	−10.2	CO2
:42	−10.5	do.
:43	−10.8	—
:43+	—	Frozen[b]

t_{sf}

−10.7°C
−10.8

July 9, 1943 G1 Distilled

Time	Temp.	Remarks
2:14	− 7.5	G1 put in
:15	− 6.5	CO2
:17	− 7.4	do.
:19	− 9.2	do.
:21	− 9.7	do.
:22	− 9.8	do.
:23	−10.1−	do.
:24	−10.2	do.
:25	−10.8	do.
:26	−11.0	do.
:27	−11.0	do.
:28	−11.3	Frozen[a]
:30	−11.0	G1 put in
:31	−10.8	CO2
:32	−11.0	—
:33	−10.9−	CO2
:34	−10.9	do.
:35	−11.0	do.
:36	−11.3	do.
:37	−11.7	Frozen[a]
:40	−10.9+	G1 put in
:41	−10.4	CO2
:42	−10.6	do.
:43	−10.9	do.
:44	−11.0	do.
:45	−11.2	do.
:46	−11.5	do.
:47	−11.7	Frozen[b]

t_{sf}

−11.3°C
−11.7
−11.7

the difference between those two freezing points is no greater than that between the two for P24, although each of the last corresponded to a chilling of 8 minutes, whereas for G5 the first chilling lasted an hour, and the second only 10 minutes. This and many other similar cases, some of which will be presently considered, indicate that the duration of chilling is of very minor importance, provided that the chilling is not too brief; i.e. that the bath is not too cold.

The sets of January 19, 1937, and July 16, 1943 (C12 and P25) are typical of those made at the beginning of the day, when the bath is initially above 0°C. It will be noticed that the observed spontaneous freezing point is the same whether one starts with the bulb and bath both above 0°C, or with the bulb above 0°C and the bath at a temperature that is only slightly above the specimen's t_{sf}.

The sets of March 24, 1937, and July 9, 1943 (C42 and G1) are typical of a large number of sets of observations on bulbs for which the t_{sf} was approximately known.

If the freezing does not occur within less than 6 or 7 minutes after the bulb is placed in the bath, the observed t_{sf} is essentially the same as if a longer time had been taken. If it occurs much quicker than that, the result is suspect, and the values found may not be reproducible.

The quicker freezing may be caused (1) by the bath being too cold, causing the water in contact with the wall of the bulb to freeze before the bulk of the liquid has become supercooled. In that case, ice grows progressively from the wall inward, the bulk of the water never becomes supercooled, and the initial thin layer of ice on the wall is crystal clear and very hard to see. Or (2) a minute star of ice may have formed on the wall in the gas phase, and set off the freezing of the supercooled water as soon as the star had grown large enough to touch the water, irrespective of the actual amount of the supercooling. A case of that kind will be discussed later. Or (3), as realized later, the t_{sf} of the specimen may be drifting upward, and from that cause may meet the falling temperature of the bath more quickly than it would otherwise have done.

On a few occasions it has seemed that the meniscus has frozen at a higher temperature than the t_{sf} of the liquid in bulk. And that something occasionally occurs on the wall of the bulb very near the meniscus is shown by the fact that occasionally a very slight and gentle pouring of the water over that region will initiate freezing at an abnormally high temperature, whereas in general the water can be poured with impunity over the entire internal area of the bulb. I have not identified that something. It may be a mote, an ice crystal, or something else. Its occurrence seems to be exceptional.

f. Reproducibility

Illustrative examples of the reproducibility of the observed values of the spontaneous-freezing-point are given in table 3.

For several of the bulbs there are numerous (5 to 12) freezings made on a single day and following one another as closely as was practical. In those 8 long series, totalling 66 freezings, there are 2 temperatures that are abnormally high as compared with their neigh-

TABLE 3

Reproducibility of the Observed Temperatures of Spontaneous Freezing

For additional information about the specimens, see appendix. The water was observed to freeze spontaneously when the temperature of the bath was $t°C$. No value of t observed during any of these intervals has been omitted from the table.

C38 Brook		P10 Vac. Dist.		CIII Vac. Dist.		C12 Dist.		C5 Dist.	
Date	t	Date	t	Date	t	Date	t	Date	t
1943		1943		1943		1936		1943	
Apr. 9	−10.8	Apr. 14	−16.1	Apr. 12	−17.2	Dec. 22	−6.8	July 6	−11.6
	−10.3		−16.0		−17.1		−6.3		−12.0
	−10.5		−16.2		−17.4		−7.0		−12.0
	−10.1		−16.1		−17.7		−7.1		−12.0
	−10.3		−12.9[a]		−17.0		−6.6		−12.1
	−10.6		−16.8		−17.6		−6.5	July 9	−12.0
	−10.6		−16.9		−17.8		−6.8		−12.1
	−10.7	Apr. 15	−16.2		−17.2		−6.9	July 16	−12.2
	−10.4		−16.6		−17.6		−6.3		−12.1
	−10.5		−16.5	Apr. 13	−17.8		−6.9		−12.2
	−10.8		−16.2		−17.7		−6.8		
	−10.6		−16.2		−17.6	1937			
			−16.0		−17.6	Jan. 5	−7.0		
			−16.2		−17.6		−6.7		
			−16.0		−17.3		−6.4		
					−17.4				
					−17.2				

C23 Snow		C40 Pool		P25 Dist.		P2 Conductivity		C10 Conductivity	
Date	t	Date	t	Date	t	Date	t	Date	t
1937		1937		1943		1937		1937	
Feb. 17	−5.8	Mar. 4	−6.0	July 5	− 9.7	Jan. 19	−13.4	Feb. 25	−16.0
	−6.0		−6.0		− 9.8		−13.3		−16.0
	−6.0	Mar. 6	−6.9[b]		− 9.8	Jan. 28	−13.8		−16.0
	−6.0		−6.0	July 6	− 9.6		−13.6	Mar. 10	−15.9
	−7.5[b]	Mar. 8	−6.0		−10.0		−13.5		−16.0
	−6.0		−6.0		− 9.5	Feb. 24	−13.4	Apr. 14	−16.0
Feb. 24	−6.1	Mar. 22	−5.6	July 9	− 9.6		−13.0[c]		−16.0
	−6.1		−6.0		− 9.9		−13.1		
Mar. 9	−6.0	Apr. 23	−6.0		− 9.7	Mar. 10	−13.2		
	−5.9		−6.0	July 16	− 9.9		−13.2		
	−6.0				−10.1				
Mar. 22	−5.9				− 9.9				
Apr. 22	−5.8								
	−5.9								

[a] P10. I cannot account for this high reading.
[b] C23, C40. The temperature of the bath when the bulb was placed in it was below the t_{sf} of the bulb.
[c] P2. The water froze within 2½ minutes after the bulb was placed in the −13.0°C bath.

bors. If those 2 are now disregarded, and each series is separately averaged and its mean deviation from that average determined, the 8 mean deviations so obtained range from 0.06 to 0.33°C, and average 0.18°C.

Whence it seems valid to assume that *for an unchanging specimen* the experimental errors in the determination of the temperature of the specimen at the instant of freezing and the chance of variation in that temperature do not together exceed about 0.2°C.

Furthermore, it will be noticed that, although the observations of certain of those specimens were extended over weeks or months, every observed value of t_{sf} for a given specimen was essentially the same, although such long intervals are favorable to the occurrence of changes in the specimen.

Whence it is obvious that, whatever irregularities may exist in certain cases, there is a firm substratum of reproducibility upon which one can base his investigations. Chaos does not reign supreme.

Furthermore, since these data give no evidence of any wide-ranging chaotic distribution of the successive determinations of t_{sf}, one must conclude, contrary to a widely held belief, that such chaos is not a fundamental characteristic of the phenomenon of spontaneous freezing.

g. Graphs

The more detailed study of the observations of the sealed bulbs to be given in the following sections is much facilitated by the use of graphs. They have been used, and will be found in the proper places. They are all constructed on the same plan, which requires some explanation; that explanation is given in this section.

In each of those graphs (figures 3 to 27, inclusive) the dots indicate the temperatures of the bath when the specimen froze spontaneously. Every such freezing made after December 19, 1939, when the unsilvered Dewar cylinders were first used, is recorded, saving only those in which the bulb was inadvertently or otherwise plunged into a bath that was plainly at a lower temperature than that corresponding to the specimen's current t_{sf}. In certain cases, a specimen would freeze quickly (within 4 or 5 minutes after it was placed in the bath), although there was no other reason for thinking that the bath was too cold; and now and again it happened that a specimen froze spontaneously at such an unanticipated, high temperature that the temperature of the bath was being lowered more rapidly than is desirable. The dots corresponding to such unusual observations are marked with an interrogation point; the true t_{sf} of the specimen at that time is certainly not lower, but may be higher, than that indicated by such a dot.

Occasionally the freezing occurred simultaneously with a mechanical disturbance of the bulb. Each corresponding dot is marked with a d. The temperature at which the specimen would at that time have frozen spontaneously is certainly not higher, but may be lower, than that indicated by such a dot.

In all cases, temperatures (°C) are indicated along a vertical linear scale; and the dots are distributed horizontally to the right in chronological order, but their spacing is not uniform with the time.

If consecutive freezings are separated in time by less than a month and if the specimen throughout that interval has either remained at room temperature or been simply melted, then the corresponding dots are, in general, always spaced horizontally by the same amount, whatever may have been the elapsed time; and they are connected by continuous lines.

If, however, two consecutive freezings are separated by an interval of at least a month, or if between them the specimen has been given some special treatment (heating, prolonged cooling, opening the bulb, etc.), or if for some other reason it seems desirable to indicate that the specimen may have changed between the two freezings, then the corresponding dots are usually separated by five times the normal spacing, and are connected by a dashed line.

When the elapsed time is to be considered, a number indicating that time is written along the dashed line. If the time is less than a month, it is expressed in days and the numeral is followed by a d. If the time equals or exceeds a month, it is expressed in months, and the numeral, rounded to the nearest month, stands alone.

Warming (heating to 54°C), heating (suspending the bulb in actively boiling water, about 97°C), prolonged cooling, and opening of the bulb are indicated, respectively, by the letters W, H, c, and O written against the dashed line. If the duration of the warming and of the heating is not given in the legend or in the accompanying text, reference should be made to the appendix.

It will be noticed that the observations are very irregularly distributed. In some cases, many observations are taken within a short period of time, sometimes within a single day, but frequently a small group of two or three observations are separated by months from their nearest neighbors. This irregularity arose in part from the exploratory nature of the work, and in part from the impression produced upon the observer by the earlier observations, which indicated a reasonable constancy in t_{sf} over short periods of time, and a slow progressive fall in t_{sf} over long ones. Numerous, closely spaced observations were taken for the purpose of establishing or checking that constancy; and it was thought that the progressive fall in t_{sf} would be satisfactorily determined, and much time would be saved for the investigation of other pertinent facts, if groups of only a few determinations each, especially if self-concordant, were made at long intervals. As will be seen, the irregularity so introduced, although justified under the circumstances attending the work, was somewhat unfortunate.

TABLE 4
INDEX TO GRAPHS

Bulb	Fig.	Bulb	Fig.	Bulb	Fig.	Bulb	Fig.
CI	8	C26	11	C49	12	P16	15, 25
CII	27	C27	3	C50	20	P17	23
CIII	5	C28	10	C51	16, 21	P18	9, 23
CIV	22	C29	13	G1	18	P20	12
C7	22	C30	13	G2	18	P21	20
C8	3	C31	13	G3	18	P22	6, 17
C9	3	C32	11	G4	18	P23	17
C10	19	C33	3	G5	18	P24	17
C11	19	C34	3	P1	19	P25	17
C12	15	C35	14	P2	19	P26	9
C13	14	C37	16	P4	11	P27	17
C14	7	C38	26	P5	10	P28	20
C15	9	C39	26	P6	11	P29	20
C17	6	C40	4	P7	15	P30	3, 20
C18	27	C41	26	P8	4	P31	20
C19	6, 27	C42	12	P10	5, 24	P32	20
C20	27	C43	11	P11	16, 24	P33	3, 20
C21	27	C45	16	P12	9, 24	P34	20
C23	4	C46	26	P13	27	P35	16, 21
C24	10	C47	10	P14	6, 25	P36	21
C25	4	C48	3	P15	25	P37	21

There are one or more graphs for each of the 84 sealed specimens, mainly so grouped as to facilitate their study, but in some cases so as to reduce the number of figures required. The number of the figure, or figures, in which the graph for a particular specimen is given may be found from either table 4 or the appendix.

h. Secular Variation

Although the data presented and studied in subsection f showed that the observed value of t_{sf} for a specimen may remain unchanged for weeks or months, such constancy over long periods is not a basic characteristic of t_{sf}. Rather, it seems that every specimen is continuously changing, and that those changes in it may be, and often are, attended by changes in the value of t_{sf}.

Those secular variations in t_{sf} may be no more than a *monotonous fall* in the temperature as the specimen ages, as illustrated by the graphs in figure 3. Similar graphs may be found in other figures, notably in figure 13.

It should be noticed that at room temperature the change may be still in progress even 5 years after the bulb was sealed. For example, at about 67 months after $C34$ was sealed its t_{sf} was about $-10.7°C$, and 46 months later it was $-12.8°C$, a fall of $2.1°C$ in 3 years and 10 months. Furthermore, although the slopes at different portions of a graph differ in significance (see subsection g), it is obvious that the changes in t_{sf} that are associated with the later and

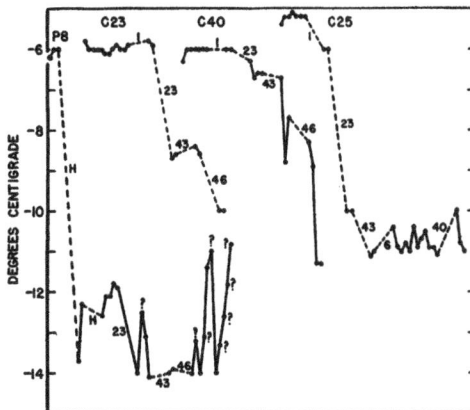

FIG. 4. Preferred temperatures: I. $P8$ cold faucet; $C23$ melted snow; $C25$ large spring; $C40$ surface of stagnant pool. Heatings (H), left to right: $P8$, 2 and 5.5 hours. See section B,I,1,g.

longer intermissions (43 and 46 months) are in general less than those associated with the earlier and shorter ones. That is, the rate at which t_{sf} falls decreases, in general, as the specimen ages. The available data do not justify a more definite statement. Seldom has this spontaneous monotonous fall been observed to extend significantly beyond $-14°C$.

Such a monotonous fall in t_{sf} may, however, be seriously disturbed by the occurrence of intervals during which the observed t_{sf} takes the same, or nearly the same, value time after time for weeks or months.

These *preferred temperatures* may occur at any stage in the life of the specimen. In some cases, every observed value throughout the interval is at the same preferred temperature; in other cases, those at the preferred temperature are interspersed with others.

Illustrations of the existence of preferred temperatures, some occurring when the specimen is very new, others when it is old, are given in figure 4. It should be noticed that the preferred temperature at $-6°C$ is common to both $C23$ and $C40$, and perhaps to the other two; whereas that of $P8$ at $-14°C$ lies much lower than any observed t_{sf} for the other three, and that of $C25$ at $-11°C$ is not clearly indicated on any of the other three, although it is suggested by two points on the graph for $P8$.

Other interesting features of such graphs, as affected by the existence of preferred temperatures, are illustrated in figure 5. It will be noticed that the observed t_{sf} for $CIII$ reached the preferred temperature $-17°C$ a little over 2 months after the bulb was sealed, approaching it from above; that value was not again observed until 66 months later, when the approach was from below. For $P10$, $-16°C$ is plainly a preferred temperature, and the lowest observed values suggest

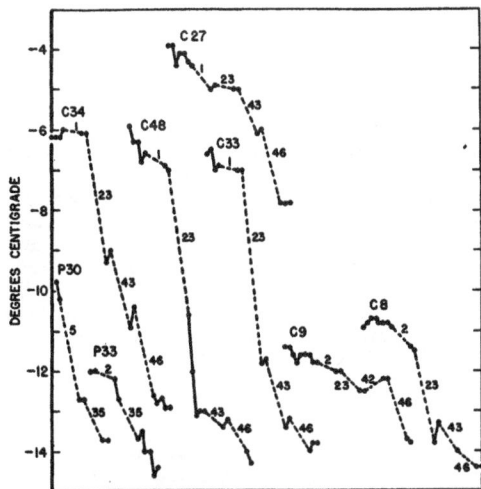

FIG. 3. Monotonous fall in t_{sf}. $C8$ and $C9$ distilled; $P30$ distilled, filtered, boiled; $P33$ redistilled after boiling for 15 hours; $C27$ stagnant pool, sediment; $C33$ and $C34$ hot faucet; $C48$ cold faucet. $P30$ and $P33$ in dusty room. See section B,I,1,g.

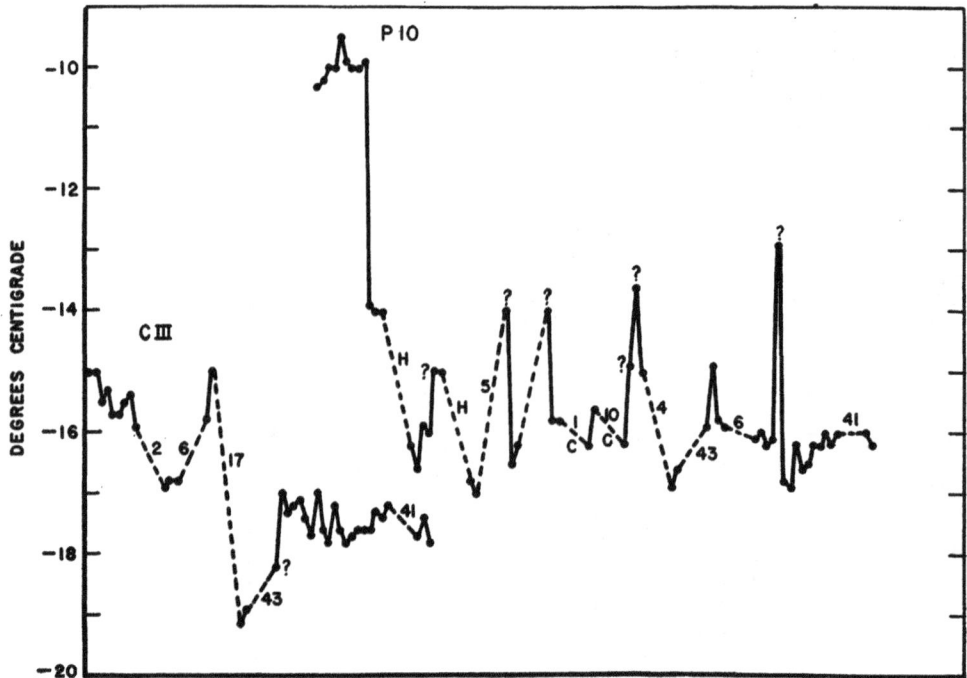

FIG. 5. Preferred temperatures: II. Vacuum distilled, air removed by boiling (CIII), by pump (P10). Heatings (H), left to right: P10, 2.2 and 4.2 hours. See section B,I,1,g.

that − 17°C (that for CIII) may be another. Furthermore, there are five points at − 15°C (CIII has 3), but only one between that and − 16°C that differs from the latter by more than 0.2°C. That is, the concentrations of points at those integral degrees are definitely greater than at any intervening temperature. Again, there are five points at − 14°C, but none between that and − 15°C. On the other hand, there are numerous points on the P10 graph between − 16 and − 17°C, and on that for CIII between − 17 and − 18°C.

Whence it appears: (1) that a preferred temperature may be approached either from below or from above, and that the direction of approach may vary from time to time; (2) that a given specimen may have several preferred temperatures, and that the observed t_{sf} may vary from one to another, and in any order, even ignoring intermediate ones; (3) that any given preferred temperature may be common to each of several specimens.

Similar conclusions may be derived from the graphs in figure 6, in which the more clearly defined preferred temperatures for the individual graphs are these: P14, − 18°C, − 20°C, probably − 16°C and − 14°C; C19,

− 12°C, perhaps − 15 and − 16°C; C17, − 16°C; P22, − 12°C, probably − 14 and − 15.5°C.

In certain cases, when special pains have been taken to reduce the number of motes contained in the water, the passages of the observed t_{sf} from one preferred temperature to another may occur so frequently that from a casual inspection of the tabulated values one may well obtain the impression that chaos rules. However, a closer inspection will reveal that certain temperatures occur again and again, and that other neighboring temperatures seldom if ever occur. Such a distribution is incompatible with chaos; it points to law and order, even though the order be far from simple. And when the data are plotted, the nature of the order becomes obvious.

Graphs of that kind are shown in figures 7 and 8, for C14 and CI, respectively. Greater pains to remove the motes were taken with these than with any of the other sealed specimens. In each case the charge was obtained by vacuum distillation without ebullition, the air having been removed by long and repeated boiling, and the bulb was washed many times with the distillate. The reservoir of the distilling system contained initially ordinary distilled water in the case of CI, and

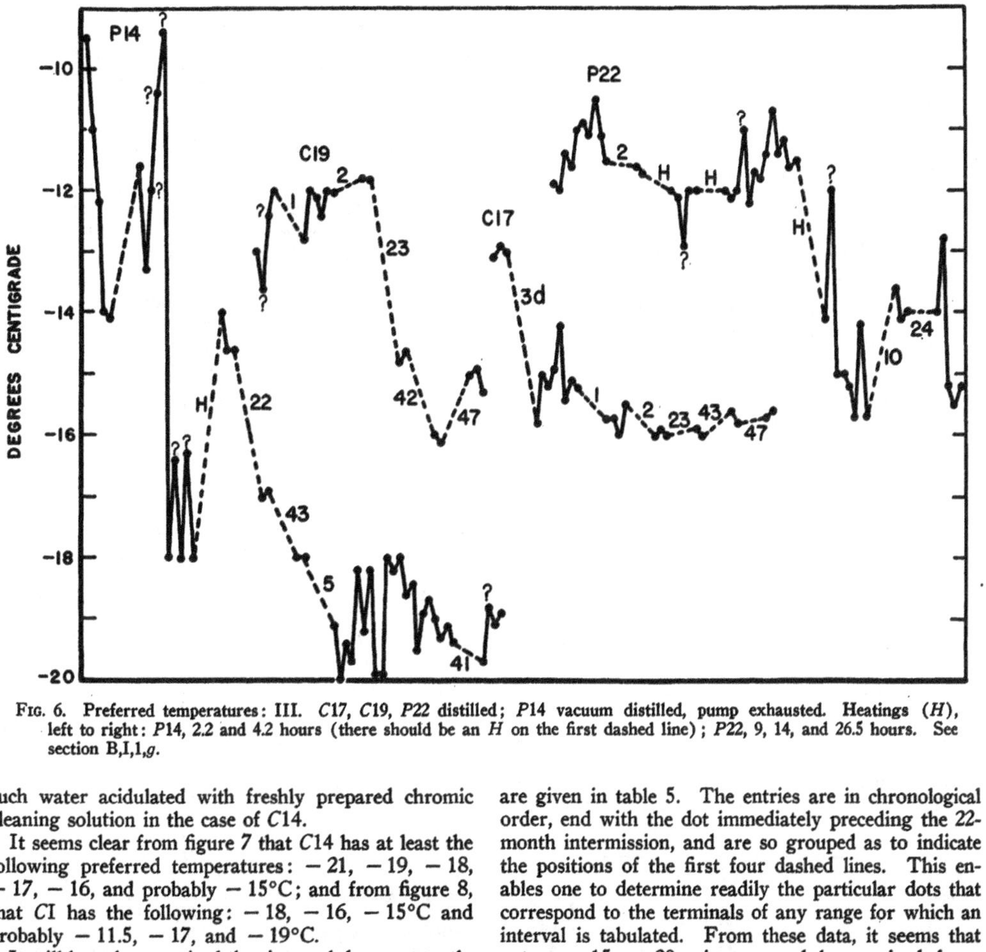

Fig. 6. Preferred temperatures: III. *C17, C19, P22* distilled; *P14* vacuum distilled, pump exhausted. Heatings (*H*), left to right: *P14*, 2.2 and 4.2 hours (there should be an *H* on the first dashed line); *P22*, 9, 14, and 26.5 hours. See section B,I,1,*g.*

such water acidulated with freshly prepared chromic cleaning solution in the case of *C14.*

It seems clear from figure 7 that *C14* has at least the following preferred temperatures: − 21, − 19, − 18, − 17, − 16, and probably − 15°C; and from figure 8, that *CI* has the following: − 18, − 16, − 15°C and probably − 11.5, − 17, and − 19°C.

It will have been noticed that integral degrees greatly predominate in the preferred temperatures to which attention has been called. I cannot now account for such predominance. There was nothing about the way the observations were taken that would lead to a predominance of integral values for t_{sf} (*cf.* table 2), and bias could not have entered, as I acquired no clear conception of the importance, or even of the existence, of such preferred temperatures until the experimental work had been concluded.

It would be most interesting to know how quickly the value of t_{sf} can change from that at the bottom of a peak to that at its top, and conversely, in such a graph as that of *C14* (fig. 7). The observations do not lend themselves satisfactorily to such a determination. However, the intervals actually involved in that graph may throw a little light on the subject. They

are given in table 5. The entries are in chronological order, end with the dot immediately preceding the 22-month intermission, and are so grouped as to indicate the positions of the first four dashed lines. This enables one to determine readily the particular dots that correspond to the terminals of any range for which an interval is tabulated. From these data, it seems that not over 15 or 20 minutes need be required for a change of 3 or 4°C. The change may be quite rapid.

These rapid changes, contrasting sharply with the very slow downward drift of t_{sf} for such specimens as those for which graphs are given in figure 3, suggest some difference in the mechanisms involved in the two cases. A possible difference is considered in the theoretical portion of this report.

In all the graphs heretofore considered t_{sf} has either progressively fallen, perhaps to what seems to be a limiting value, or has undergone striking fluctuations in value, or has exhibited a combination of the two. Only in the case of *P8* (fig. 4) is there any suggestion that t_{sf} may rise progressively. That, however, is possible. Such progressive rise is shown most strikingly by *P12* (fig. 9) and *C10* (fig. 19), and less so by *P23* (fig. 17). All of these occurred after t_{sf} had dropped

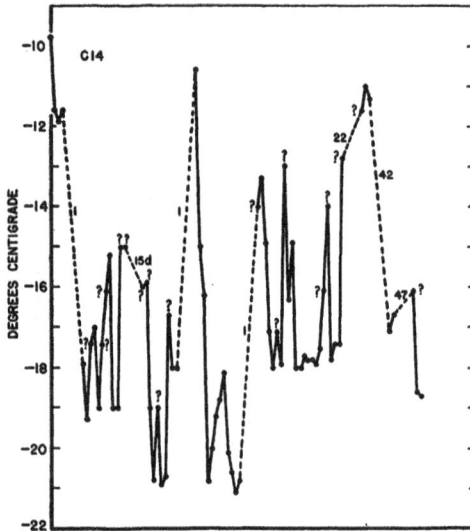

FIG. 7. Multiple preferred temperatures: I. C14 vacuum distilled from distilled water treated with chromic solution, air removed by boiling. See section B,I,1,g.

to a low value. A rise may, however, occur soon after a specimen has been sealed. In such cases it is generally short-lived, and its progress is irregular. See P22 (fig. 6), P12 (fig. 9), P5, C47, and C24 (fig. 10), P16 (fig. 15), C37 (fig. 16), P36 (fig. 21), and CIV (fig. 22).

Certain miscellaneous graphs are given in figures 9 to 12; others in figures 13 to 27, to be found in fol-

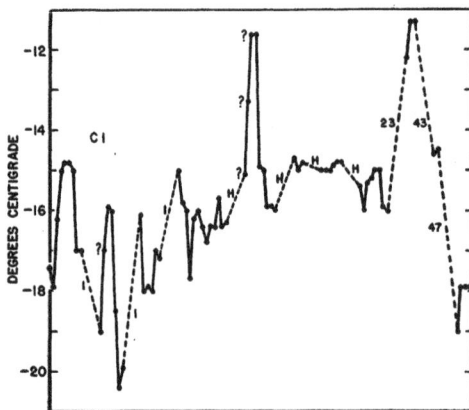

FIG. 8. Multiple preferred temperatures: II. CI vacuum distilled, air removed by boiling. Heatings (H), left to right, 0.5, 0.3, 0.5, and 2 hours. See section B,I,1,g.

lowing subsections. They, as well as those already given, are pertinent to the study. Taken all together, they display, as stated in subsection g, every pertinent value of t_{sf} obtained in the study of the sealed bulbs listed in the appendix.

i. Prolonged Chilling

Much work published prior to this study had been done under the impression that the freezing or nonfreezing of a specimen of water depended upon how long it had been kept at the subzero temperature. Hence, although even the preliminary work in 1936 indicated that the freezing or nonfreezing depended primarily upon the temperature to which the specimen

TABLE 5

TIME INTERVALS: C14, FIG. 7

Under "interval" is tabulated the time that elapsed between the extreme temperatures recorded under "range"; under Δ is that range, in °C. The breaks in the table indicate, respectively, the positions of the first 4 dashed lines in the graph in figure 7.

Temperatures are expressed in °C

Temperature		Interval		Temperature		Interval	
Range	Δ	da.	min.	Range	Δ	da.	min.
− 9.8 to −11.9	2.1	—	117	−15.0 to −16.2	1.2	—	18
				−16.2 to −20.8	4.6	—	130
−19.3 to −17.0	2.3	—	14	−20.8 to −19.2	1.6	—	16
−17.0 to −19.0	2.0	5	—	−19.2 to −18.8	0.4	1	—
−19.0 to −16.1	2.9	—	12	−18.8 to −18.1	0.7	—	8
−16.1 to −15.2	0.9	11	—	−18.1 to −21.1	3.0	—	43
−15.2 to −19.0	3.8	—	41				
−19.0 to −15.0	4.0	12	—	−13.3 to −18.0	4.7	—	47
				−17.9 to −13.0	4.9	¼	—
−15.9 to −20.8	4.9	—	56	−13.0 to −16.3	3.3	—	20
−20.8 to −19.0	1.8	—	6	−16.3 to −14.9	1.4	—	9
−19.0 to −20.9	1.9	—	15	−14.9 to −18.0	3.1	—	19
−20.9 to −20.7	0.2	—	12	−18.0 to −17.9	0.1	2+	—
−20.7 to −16.7	4.0	13	—	−17.9 to −17.5	0.4	1+	—
−16.7 to −18.0	1.3	—	21	−17.5 to −14.0	3.5	—	13
				−14.0 to −17.8	3.8	—	31
−10.6 to −15.0	4.4	¾	—	−17.4 to −12.8	4.6	15	—

had been cooled, and only very slightly, if at all, upon how long the specimen had been kept at a subzero temperature, it seemed desirable to keep some of the sealed bulbs continuously supercooled for a prolonged period. Such tests, extending over 30 days, during which certain of the sealed bulbs were kept continuously between − 3 and − 12°C without freezing, were reported in the paper of 1938.

Since then, a much longer test has been made with the four bulbs, C11, C35, C49, and P10. These bulbs were suspended at first in alcohol, and later in a strong solution of salt, in a vessel placed in the ice-cube compartment of an electrical refrigerator. (The salt solution was brought to the same temperature as the alcohol before the bulbs were transferred from the latter to the former, the change having been made because of the excessive evaporation of the alcohol.) At first, the temperature of the bath was read several times a day. Its variation during a day having been found to be less than a degree, readings were reduced to two a day. Bulb $C49$ was a relatively new one, having been sealed only two months before the test began.

Those bulbs were placed in a cold bath on December 14, 1937, and were removed for freezing on January 19, 1938; for those 36 days their temperature had been continuously about − 6°C. Immediately after that freezing, they were returned to the cold bath, and during the next 4 days the thermostat was adjusted for a lower temperature; then for the next 312 days (until

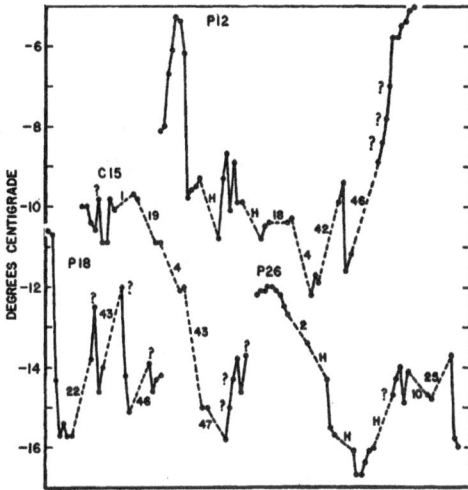

FIG. 9. Miscellaneous secular variations: I. $C15$ melted snow; $P12$ vacuum distilled, pump exhausted; $P26$ distilled (special); $P18$ residue from distillation. Heatings (H), left to right: $P12$, 2.2 and 4.2 hours; $P26$, 9, 14, and 26.5 hours. See section B,I,1,g.

December 2) their temperature lay continuously between − 8.0 and − 10.3°C. The new bulb ($C49$) froze 4 times before February 12, but after that, it did not freeze again during the test. None of the other three froze at all.

On December 2, the bulbs were transferred one by one from the brine to alcohol at − 8°C, and from that to the freezing bath adjusted to − 8 or − 9°C when the bulb was placed in it, and then cooled to freezing. The results may be found in table 6. It will be noticed that in each case, excepting $C49$, the spontaneous freezing point after the prolonged chilling was essentially the same as that before it; and for $C49$ it was slightly

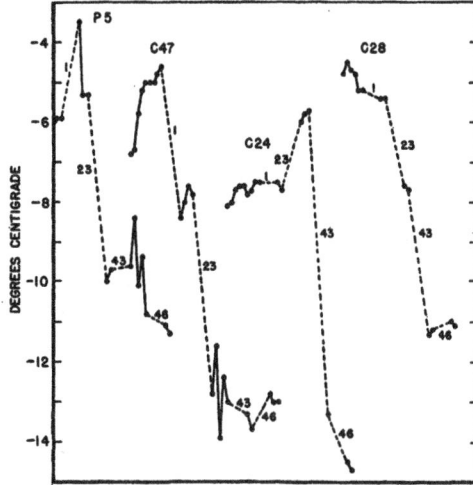

FIG. 10. Miscellaneous secular variations: II. $P5$ stagnant pool; $C47$ cold faucet; $C24$ melted snow; $C28$ river. See section B,I,1,g.

lower, which is what one might have expected for such a relatively new bulb.

Hence one may conclude that the duration of the chilling is of no consequence other than as it affects the aging of the specimen.

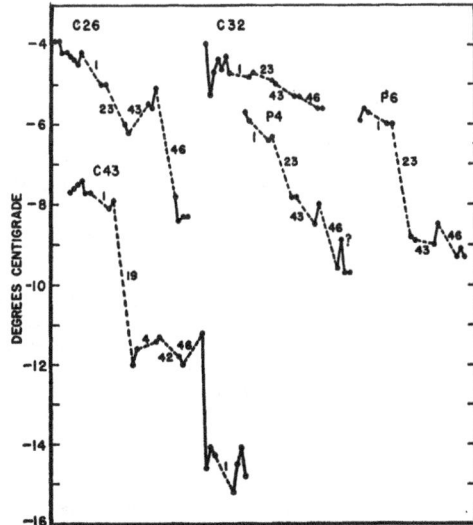

FIG. 11. Miscellaneous secular variations: III. $C26$ small, alga-covered spring; $C32$ melted snow; $C43$ aquarium; $P4$ brook; $P6$ mixture of brook with water from cold faucet. See section B,I,1,g.

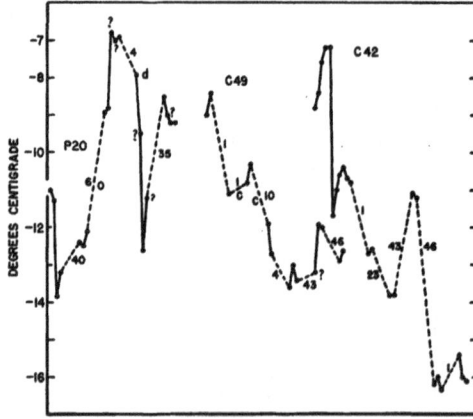

FIG. 12. Miscellaneous secular variations: IV. *C42* aquarium; *C49* old stock of distilled; *P20* from steamer for bulbs, at *O* neck shortened, opened, and resealed in dusty room. See section B,I,1,*g*.

j. Preheating

Two distinct effects may be produced by preheating a melt: (1) A direct, immediate change, and (2) an after-effect, modifying the subsequent behavior of the melt.

The data obtained in this study do not suffice to distinguish at all clearly such after-effects from the unpredictable irregularities that have already been discussed. However, it seems not unreasonable to suppose that, in general, heating does little more than hasten the changes that occur more slowly as the specimen ages at room temperature. That seems to be indicated by the graphs shown in figure 13. *C29* and *C30* were each charged with water drawn through iron pipes from the same wooden storage tank. The sole difference is that a portion of the pipe through which the water in *C30* passed was heated in a water-back stove and led to an iron hot-water tank, from which the water charging *C30* was drawn; whereas the charge in *C29* had never been heated. The bulbs were charged in quick succession, were promptly sealed, and freezings were made the next day. Although the initial values of their t_{sf} differed widely, after 9.5 years the two specimens froze at the same temperature. *C31* was charged at the same time as the others, and differs from *C29* solely in having been charged directly from the iron pump which supplied the storage tank.

Another important inference may be drawn from these graphs; viz. that the slow fall in the value of t_{sf} for *C29*, and consequently in other similar cases, does not result from a slow solution of the glass, since the same fall was quickly produced by heating the water in iron, and an equally long standing of the heated water in glass led to only a slight additional fall.

The direct, immediate change in t_{sf} produced by preheating seems at first sight to be determinable from the difference between the values of t_{sf} before and after heating. However, things are not so simple. Between the heating and the subsequent freezing, the melt has been for an appreciable time at temperatures below that to which it was heated. The difference between the two observed values of t_{sf} is determined by that combination of heating and cooling, and as there were appreciable intervals of time (often a day or more) between each determination of t_{sf} and the heating, irregularities of various kinds may affect one's conclusions. Furthermore, the heatings for intervals longer than 4 hours consisted of a series of heatings separated by coolings of one or more hours duration, often over-

TABLE 6

PROLONGED CHILLING OF NO EFFECT

Every freezing of each of these bulbs from November 16, 1937, to April 7, 1939, is accounted for in this table. When freezing occurred in less than 6 or 7 minutes, the temperature of the bath is given in the footnotes at 1-minute intervals, the first being the temperature when the bulb was placed in it, and the last that at which the water froze. The graphs may be found in the indicated figures.

Temperature of spontaneous freezing

Bulb	C11	C35	C49	P10
Fig.	19	14	12	5
Water	Conductivity	Cold tap Heated	Distilled	Vac. dist. Pump exh.
Sealed	12-10-36	3-10-37	11-5-37	3-10-37
	°C	°C	°C	°C
1937				
Nov. 16	−15.7	−14.1	− 9.0	*a*
Nov. 16	−15.7	−13.7	− 8.4	−16.5
Nov. 16	—	—	—	−16.2
Dec. 6, 7	—	—	*b*	—
Dec. 11	−15.3	−13.4	*c*	*d*
Dec. 11	−15.2	−13.4	−11.1	−15.8
Dec. 11	—	—	—	−15.8
Dec. 14	—	—	*e*	*e*
1938				
Jan. 19	−15.1	−12.8	−10.9	−16.2
Jan. 19	−15.2	−13.0	−10.3	−15.6
Jan. 24+	*f*	*f*	*f*	*f*
Dec. 2	−15.3	−13.6	−11.9	−16.2
Dec. 2	−15.6	−13.3	−12.4	*g*
Dec. 2	—	—	—	−15.0
1939				
Apr. 4–7	−15.3	−13.2	−13.6	−16.9
Apr. 4–7	−15.3	−13.1	−13.0	−16.6
Apr. 4–7	—	—	−13.4	—

a Quick freeze, −15.1, −14.0°C frozen.
b Kept frozen for 4 hours on each day.
c Quick freeze, bath too cold (−14.4°C).
d Bath at −14.0°C, froze in less than 1 minute.
e Between −5 and −7°C for 36 days (Dec. 14 to Jan. 19).
f Continuously −8.0 to −10.3°C for 312 days.
g Two quick freezings: −15.2, −14.5, −14.4, −14.9°C frozen; −14.2, −13.2, −13.6°C frozen.

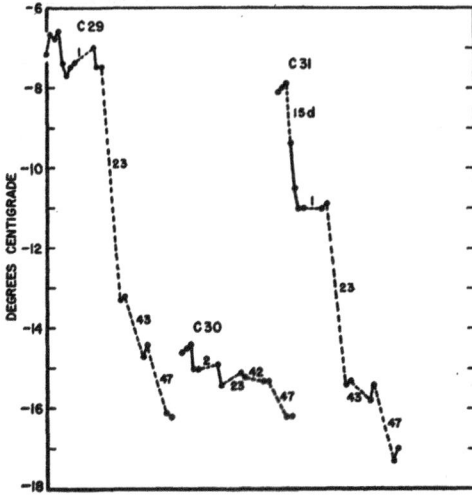

FIG. 13. Effect of heating: I. Well-water. *C29* cold faucet; *C30* hot faucet; *C31* direct from pump. See section B,I,1,*g*.

night. All these departures from ideal simplicity should be remembered.

Nevertheless, the actual data do exhibit certain features which are worthy of notice, and which may with some confidence be taken as indications of changes produced directly by the heat treatment, among which is the striking contrast between the graphs, already noticed, for *C29* and *C30* (fig. 13).

The type of effect that is produced by heating a sealed specimen is illustrated by the graphs in figure 14. A mere warming (*W*) to 54°C for 8 hours produces little, if any, change in t_{sf}. In contrast to that, the first heating (*H*) in boiling water for as short a period as 2 hours may cause a marked fall in the temperature of spontaneous freezing. The effect, if any, of a subsequent heating is so slight that it is not certainly distinguishable from irregularities that might have occurred had the specimen not been heated. The graphs in figure 15 lead to the same conclusion; as do also those in figure 16.

In figure 16, however, it will be noticed that even the first heating (*H*) has little, if any, effect upon the subsequent values of t_{sf} for *C37*, *P11*, and *P35*; and that its apparent effect on t_{sf} for *C51* may be entirely spurious, arising from variations that would have occurred had the specimen not been heated. Why did heating produce no effect in these cases? An examination of the 10 graphs displayed in figures 14, 15, and 16 reveals that the temperature to which t_{sf} was lowered by the first heating (*H*) varied from − 11 to − 14, (perhaps) − 15°C, depending upon the specimen, and that the value of t_{sf} immediately preceding the heating of *C37*, *P11*, *P35*, and *C51* lay within that range of tempera-

ture. Furthermore, as previously observed, heatings subsequent to the first have little if any effect on the following values of t_{sf}.

From all of which, it seems fair to conclude, in the absence of adverse evidence from other graphs for heated specimens (fig. 4, 5, 6, 8, 9, 17, 18, 19, 21, 24, 25), that the effect of an initial heating (*H*) of 2 hours or more brings the specimens studied in this work to such a thermally indifferent state that they are not affected by further heatings or by prolonging the heating (*C12*, fig. 15, was heated for 18.5 hours); and if a specimen be initially in that indifferent state, then heating does not affect its t_{sf}. In this work the value of t_{sf} when a specimen is in its thermally indifferent state lies below a certain temperature lying between − 11 and − 15°C depending on the specimen. Whether a specimen ceases to be thermally indifferent when its t_{sf} rises later to a value above that region, as in the case of *P12* (fig. 9), cannot be determined from the present data.

In subsection *h* it was remarked that seldom has a spontaneous monotonous fall in t_{sf} been observed to extend significantly beyond − 14°C. Whence it seems probable that the monotonous change accompanying the aging of a specimen is closely related to—is perhaps

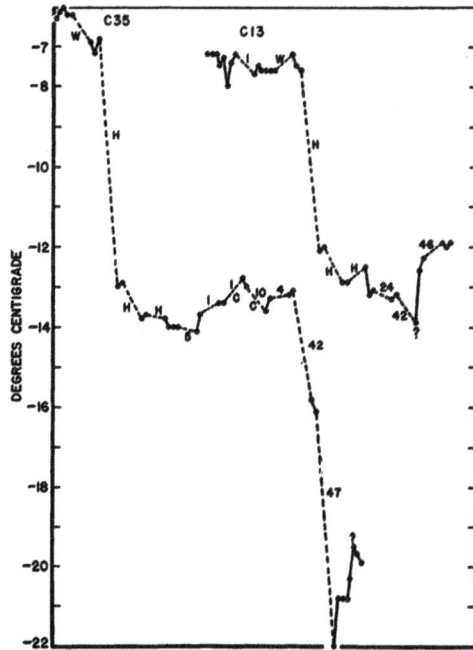

FIG. 14. Effect of heating: II. *C35* cold faucet; *C13* distilled, covered with castor oil. In each case, warmed (*W*) 54°C for 8 hours, heated (*H*), left to right, for 2, 2, and 5.5 hours. See section B,I,1,*g*.

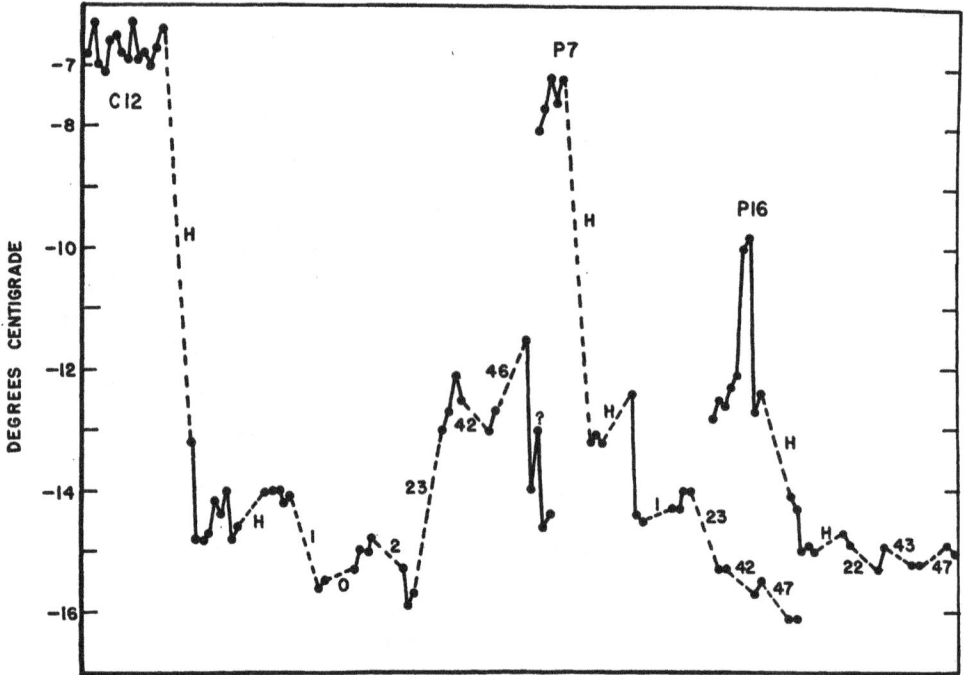

FIG. 15. Effect of heating: III. *C*12 distilled, at *O* opened for 5 days; *P*7 melted snow; *P*16 vacuum distilled, pump exhausted. Heatings (*H*), left to right: *C*12, 18.5 and 17.7 hours; *P*7, 2 and 5.5 hours; *P*16, 2.2 and 4.2 hours. See section B,I,1,*g*.

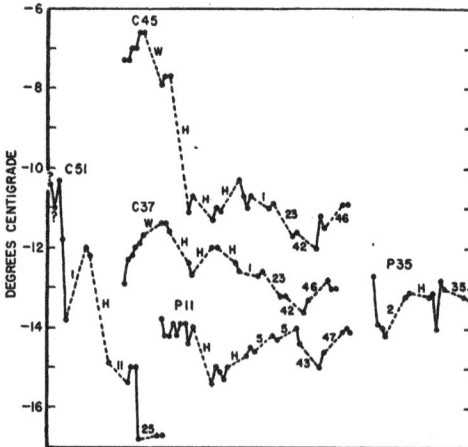

FIG. 16. Effect of heating: IV. *C*45 melted manufactured ice; *C*51 distilled from well-water, done in dusty room; *P*35 well-water, charged in dusty room; *C*37 from steamer for bulbs; *P*11 residue from distillation. Warmed (*W*) *C*45 and *C*37 54°C for 8 hours. Heatings (*H*), left to right: *C*45 and *C*37, 2, 2, and 5.5 hours; *C*51 and *P*35, 5.5 hours; *P*11, 2.2 and 4.2 hours. See section B,I,1,*g*.

identical with—that produced by heating the specimen to 97°C. In terms of the suggestion advanced in the early paper [28], each results from a solution of motes suspended in the water. Solution is hastened by heating.

But why does the monotonous fall in t_{sf} cease somewhere in the range − 11 to − 15°C, and what determines the actual temperature at which it stops for a given specimen?

An examination of the graphs reveals that the monotonous fall of t_{sf} accompanying the aging of a specimen, as well as the abrupt fall caused by the first heating, ceases at or near a preferred temperature of the specimen. Whence it may tentatively be assumed that the termination of the fall and the existence of a preferred temperature at that place are intimately related. A discussion of that relationship is deferred to subsection *g* of the section dealing with the New Theory that is herein proposed.

k. Classified Waters

In this section the data are arranged, and in part discussed, with reference to the source or treatment of the water to which each group refers. Thirty-six additional graphs are given in 11 figures; for the other 48, reference is made to the preceding figures. The waters are classified under the following heads: (1) Special specimens, (2) Conductivity water, (3) Ordinary distilled water, (4) Vacuum distilled water, (5) Other distilled waters, (6) Distillates and residues, (7) Washington City water, (8) Aquarium waters, (9) Snow and ice, (10) D. C. surface waters, (11) Maryland waters (surface and subterranean), and (12) Mixtures of water and alcohol. At the end of this subsection, the extreme observed values of t_{sf} for each specimen are given in table 7, and certain conclusions are drawn.

(1) *Special Specimens.*—Late in the work it seemed desirable to have some bulbs cleaned, charged, and sealed by another than myself, and to have them charged directly from the Bureau's still, the distilled water in those charged by myself having been taken, by means of a pipette, from the mid-layers of the water in a stock-bottle filled now and again from the distilled water pipe-line to a laboratory. On my mentioning this to Dr. Edward Wichers, he very kindly offered to arrange with Mr. F. W. Schwab, a chemist in the Bureau, to prepare such specimens for me. Six Pyrex bulbs (*P22*

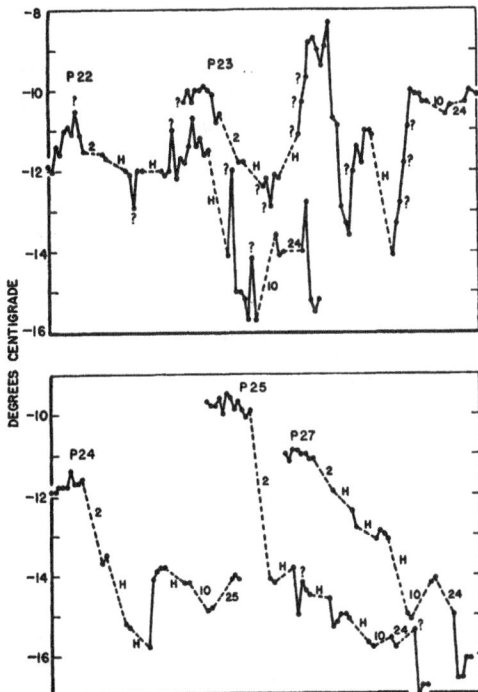

FIG. 18. Special specimens: II. Charged directly from an all-quartz still. Heatings (*H*), left to right, same for all, 9, 14, and 26.5 hours. See section B,I,1,*g*.

to *P27*) were cleaned as usual, and then given to Mr. Schwab with a request that they be recleaned with all the care a chemist would use, be charged directly from the still, and be sealed as soon as charged. He not only did that, but he prepared in addition five other similar Pyrex bulbs (*G1* to *G5*), which he charged directly from an all-quartz still, charged directly to him for this assistance. I am greatly indebted to him for this assistance.

Those eleven specimens will be called "special specimens." They were prepared in basement rooms having painted concrete floors covered in large part with rubber mats. Graphs for these specimens, with the exception of *P26*, are given in figures 17 and 18; that for *P26*, in figure 9. No striking difference between these graphs and those for the specimens previously studied is evident, but those in figure 18, for water from the quartz still, differ from one another less than do those in figure 17.

A comparison of the graphs for *P22* and *P26* with others in figures 6 and 9 fails to disclose any significant difference or superiority of *P22* and *P26* over specimens prepared by myself.

An ultra-microscopic examination of these special specimens was made for Mr. Schwab by Dr. C. P. Saylor, also of this Bureau. He arranged them in the

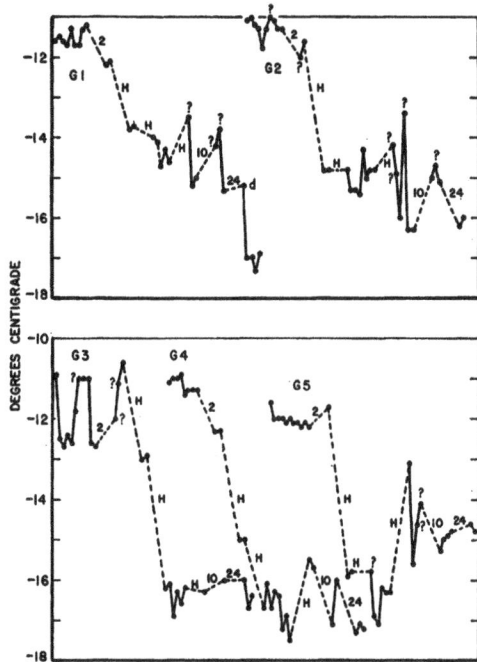

FIG. 17. Special specimens: I. Charged directly from the Bureau's still; for *P26*, see fig. 9. Heatings (*H*), left to right: *P22*, *P24*, and *P27*, 9, 14, and 26.5 hours; *P23* and *P25*, 3, 14, and 26.5 hours. See section B,I,1,*g*.

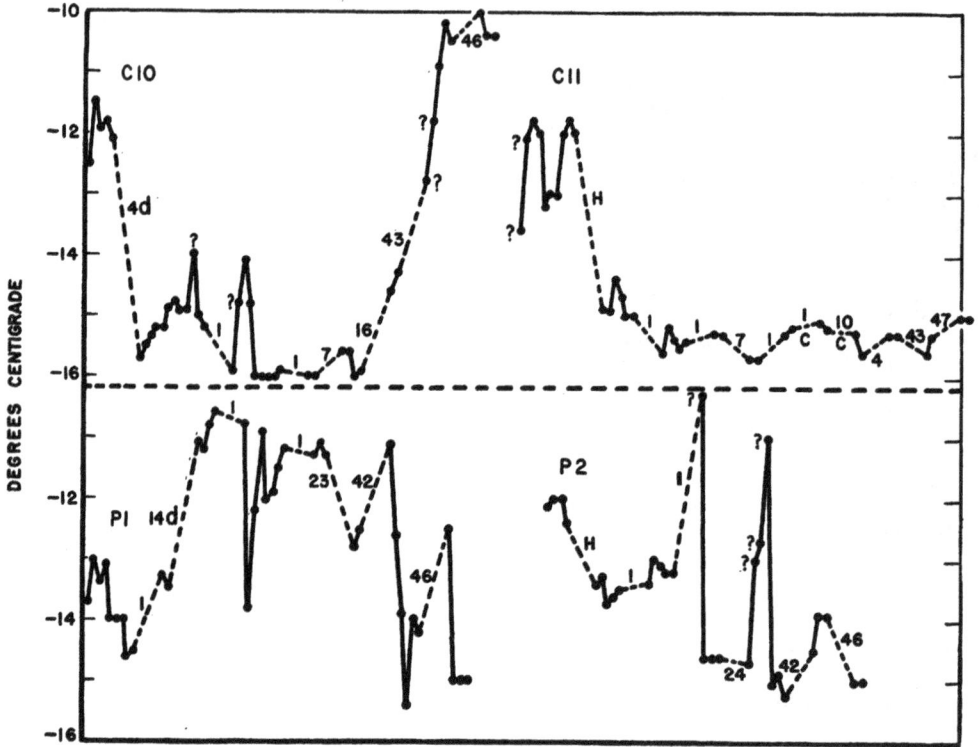

FIG. 19. Conductivity water. Charged directly from special still, specific conductivity 0.084×10^{-6} (ohm-cm)$^{-1}$ at 23°C. Heated (H) $C11$ and $P2$ for 17.8 hours. See section B,I,1,g.

following order of decreasing cleanliness: $G3$ very clear; $P23$ and $P24$ clear but noticeably less so than $G3$; in $G2$ the particles are more numerous; $G5$ and $G4$ are like $G2$; then comes $G1$; $P25$ has many more particles, mostly fairly small; $P27$ is definitely dirtier than $P25$; $P26$ dirtier than $P27$; and $P22$ contains extremely many particles of various sizes.

The failure of the graphs to suggest such an ordering of these specimens perhaps arises from the effective motes being too small to be detectable by the optical means employed. A discussion of a probable upper limit to the size of an effective mote is given in the theoretical portion of this paper.

(2) *Conductivity Water.*—Early in the work, questions arose regarding the possible effect of minute amounts of solute upon the value of t_{sf}. Dr. D. N. Craig of this Bureau very kindly charged four carefully cleaned bulbs ($C10$, $C11$, $P1$, and $P2$) with conductivity water taken directly from his still, first washing the bulbs thoroughly with the same water (specific conductivity, 0.084×10^{-6} (ohm-cm)$^{-1}$ at 23°C). Graphs for these bulbs are in figure 19. It will be

noticed that they are neither more uniform nor significantly different from those for specimens that are much less pure, e.g. the special specimens (figs. 17 and 18). The striking rise in the value of t_{sf} for $C10$ is matched by those of CIV (fig. 22) and $P12$ (fig. 24), bulbs charged with water for which no claim of high purity can be made.

(3) *Ordinary Distilled Water.*—The bulbs that were charged with ordinary distilled water taken from the laboratory stock-bottle, and the extreme observed values of t_{sf} for each, are as follows: $C8$ (-10.7 to $-14.4°C$) and $C9$ (-11.4 to $-13.8°C$), figure 3; $C17$ (-12.9 to $-16.0°C$) and $C19$ (-11.8 to $-16.1°C$), figure 6; $C49$ (-8.4 to $-13.6°C$), figure 12; $C13$ (-7.2 to $-13.9°C$), figure 14; $C12$ (-6.3 to $-15.9°C$), figure 15; and $P30$ (-9.8 to $-13.7°C$), figure 20.

$C8$ and $C9$ are duplicates; $C12$ and $C13$ were charged at the same time and from the same water, but the water in $C13$ is covered with a layer of castor oil, and that in $C12$ is not. Furthermore, a platinum wire about 1 cm long is sealed near its middle through the wall of $C12$ about 4 mm above the meniscus. It was thought that the wire might facilitate the freezing, but it produced no observable effect. It should be noticed that leaving the neck of C12 open for 5 days produced no change in the value of t_{sf} (near $-15.5°C$). During that interval the bulb was kept in a closed book case; all was done in the clean room occupied prior to 1942. The castor oil in C13 was intended to seal off the water from its vapor, and to prevent the transmission of freezing along the walls. It seems to have produced no certain effect of any kind.

C19 was charged a month after C17, but probably from the same stock; the notes are not clear. C49 was charged directly from a small Pyrex wash-bottle that had remained undisturbed for 3 months. P30 was prepared in the dusty room, and was charged with distilled water that had been boiled in an open beaker in that room, the same as was placed in the reservoir from which the charge of P29 (fig. 20) was distilled.

(4) *Vacuum Distilled Water.*—Ordinary distilled water was placed in a 100-cc Pyrex reservoir (unless another is indicated) to which the bulb to be charged was sealed, the pressure was reduced, and the system was closed. Then water was distilled into the bulb at a low temperature and without ebullition; the walls of the bulb were washed with that distillate, which was then poured back into the reservoir, all without opening the system. That procedure was repeated a number of times; then the charge was similarly distilled into the bulb, and the bulb was hermetically sealed off from the reservoir.

Three means were used for reducing the pressure: (a) The air was removed by repeated boilings in the usual manner; then, after the system had become half emptied by boiling, it was hermetically sealed while steam was still escaping briskly. (b) The air was removed by means of an oil pump, a vapor trap being inserted between the pump and the system to be exhausted. The pressure beyond the trap was reduced to 1 mm-Hg or less. The set-up was a crude make-shift; the reservoir was small (a P-bulb) and connection with the pump was by means of heavy rubber tubing. (c) The air was removed by means of the low-pressure system of the laboratory (about 10 to 15 cm-Hg).

The bulbs containing such specimens, grouped according to the method of exhaustion, are as follows: (a) Air removed by boiling. CI (-11.3 to $-20.4°C$), figure 8; CIII (-15.0 to $-19.1°C$), figure 5; CIV (-12.2 to $-20.1°C$), figure 22; C14 (-9.8 to $-21.1°C$), figure 7; P21 (-12.1 to $-15.8°C$), P29 (-7.6 to $-11.3°C$), and P31 (-7.8 to $-14.0°C$), figure 20. P30 (-9.8 to $-13.7°C$), figure 20, is charged with some of the boiled water placed in the reservoir from which the charge in P29 was distilled.

It will be noticed that for each C-bulb the graph is very jagged, and that in every case t_{sf} occasionally falls to a very low temperature (-19 to $-21°C$).

On the other hand, the graphs for the P-bulbs (which

were all charged in the dusty room and have been under observation for only slightly over 3 years) are much more simple, and in no case does t_{sf} fall as low as $-16°C$. I am inclined to think that this marked difference in behavior is to be explained by a much greater mote-content of the water used for the P-bulbs, and by the more extensive soiling of those bulbs by air-borne motes.

(b) Pump exhausted. P10 (-9.5 to $-17.0°C$) and P12 (-5.0 to $-12.2°C$), figure 24; P14 (-9.4 to $-20.0°C$) and P16 (-9.8 to $-15.3°C$), figure 25; and P17 (-9.8 to $-13.7°C$), figure 23. The graphs for the first 3 (P10, P12, and P14) are quite jagged, suggestive of those for the C-bulbs in the preceding list, but only that for P14 reaches nearly so low a temperature; those for the other two are more simple in form. The cause of these differences is not known. They may have to do with either the smallness of the reservoir used or the crudeness of the set-up, or with both.

(c) Laboratory low pressure. The only specimen in this group is P32 (fig. 20, -13 to $-16°C$), which was prepared in the dusty room, and has been studied for only 39 months. The object of this experiment was to see whether such filtering of the vapor through

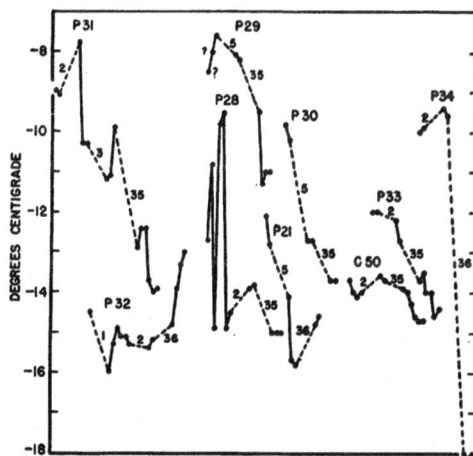

Fig. 20. Filtered, boiled, and distilled waters. All prepared in the dusty room. C50 and P28 redistilled from distilled water that had been boiled 15 hours in Pyrex, reflux condenser, collected in Pyrex; bulb charged a week later, in dust-proof box, from midlayers of that stock. P21 and P29 vacuum distilled from distilled water that had passed a German fritted filter "G19" and then been boiled; pump exhausted. P30 the preceding filtered and boiled water. P31 vacuum distilled from distilled water that had passed a Pyrex fritted filter "F"; air removed by boiling. P32 distilled under reduced pressure (10 to 20 cm-Hg) from preceding filtered water. P33 and P34 charged directly from Pyrex still containing distilled water preboiled for 15 hours in Pyrex, reflux condenser. See section B,I,1,g.

air would appreciably reduce the carrying over of large motes to the bulb being charged. Its graph does start at a much lower temperature than do those for any of the other P-bulbs in groups a and b, but at the end of 39 months t_{sf} was higher than at the start, and was rising. More work must be done before the question posed can be answered.

The very jagged nature of several of the graphs for vacuum distilled specimens is the most striking difference between the graphs for them and those .for the other specimens. Seemingly, the more successful the vacuum distillation, the more jagged is the graph.

(5) *Other Distilled Waters.*—$C50$ (-13.6 to $-14.7°C$), $P28$ (-9.5 to $-15.0°C$), $P33$ (-12.0 to $-14.6°C$), and $P34$ (-9.4 to $-18.1°C$), figure 20, were all charged with ordinary distilled water that had been boiled 15 hours in Pyrex (reflux condenser) and then distilled from Pyrex. $P33$ and $P34$ were

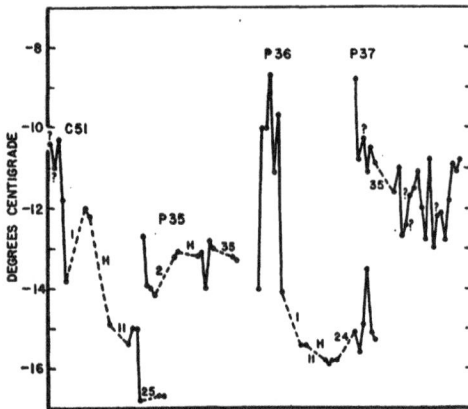

FIG. 21. Well-water (dusty room). Water was from a glass-stoppered Pyrex bottle. $P35$ charged from midlayers. The water was then distilled and redistilled in Pyrex, and the other 3 bulbs were charged directly from the still; $P36$ singly distilled, $C51$ doubly distilled, and $P37$ triply distilled. See section B,I,1,g.

charged directly from the still; $C50$ and $P28$ from some of the same water that had been bottled, and inside a nominally dust-proof box. The work was done in the dusty room.

$C51$ (-10.3 to $-16.8°C$), $P35$ (-12.7 to $-14.2°C$), $P36$ (-8.7 to $-15.9°C$), and $P37$ (-8.8 to $-13.0°C$), figure 21, were all charged with well-water (*cf.* fig. 13); $P35$ contained the raw water; $P36$ was charged with the distilled water; $C51$ with the doubly distilled water; that doubly distilled water was then boiled 7 hours with a reflux condenser, and was then distilled. $P37$ was charged with that triply distilled water. In each case, charging was directly from the still. The work was done in the dusty room. It

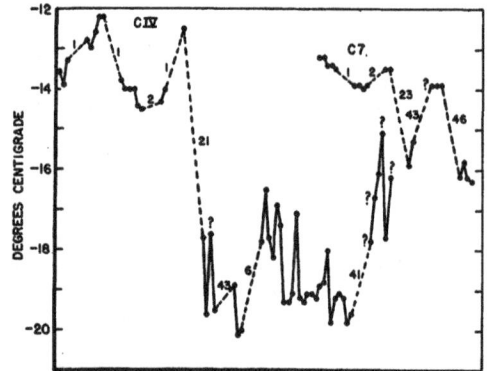

FIG. 22. Distillate and its residue (boiled out). CIV vacuum distilled, air boiled out; after charging, air was admitted, CIV sealed off, and part of residue was placed in $C7$. See section B,I,1,g.

seems that high-temperature motes, probably air-borne, were added during the treatment of the water. Nevertheless, the graph for the triply distilled water is distinctly more jagged than any of the others.

(6) *Distillates and Residues.*—In figures 20 and 22 to 25, inclusive, are shown the graphs for certain distillates and for the residues from those distillations; and one graph each in figures 12 and 16 is for a specimen from a volume of water that presumably contains an unusual number of motes and is comparable with the residues. The graphs in figure 20 have just been discussed.

Figure 22: The charge in CIV (-12.2 to $-20.1°C$) was vacuum distilled, as just described, from about 50 cc of distilled water contained in a 100-cc reservoir. Air was then admitted to the system; CIV was sealed

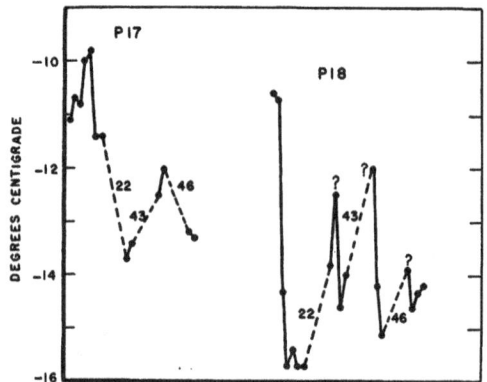

FIG. 23. Distillate and its residue (pumped out). $P17$ vacuum distilled from $P18$, pump exhausted; $P18$ residue. See section B,I,1,g.

off; and some of the residue was transferred from the reservoir to $C7$ (-13.2 to $-16.3°C$) which was then sealed. At first, t_{sf} was nearly the same for each, but 26 months later CIV froze at $-17.7°C$ and $C7$ at $-15.9°C$. However, after 9.6 years, each again froze at nearly the same temperature ($-16°C$), but in the meantime t_{sf} had been as low as $-20°C$ for CIV, although never below $-16.3°C$ for $C7$.

Figure 23: Although $P17$ (-9.8 to $-13.7°C$) was charged by vacuum distillation from $P18$ (-10.6 to $-15.7°C$), the graph for the latter (the residue) lies lower than that for the distillate.

Figure 24: $P10$ (-9.5 to $-17.0°C$) and $P12$ (-5.0 to $-12.2°C$) were each charged, and in that order, by vacuum distillation from $P11$ (-13.8 to $-15.4°C$), to which they successively were sealed. Each graph is jagged, but that for $P12$ lies much higher than does that for $P10$. The graph for $P11$, the residue, is less jagged than the others; it lies far below the lowest point of that for $P12$, and low enough to become confused with that for $P10$, if placed in the same chart.

Figure 25: $P14$ (-9.4 to $-20.0°C$) and $P16$ (-9.8 to $-15.3°C$) were each charged, but in reverse order,

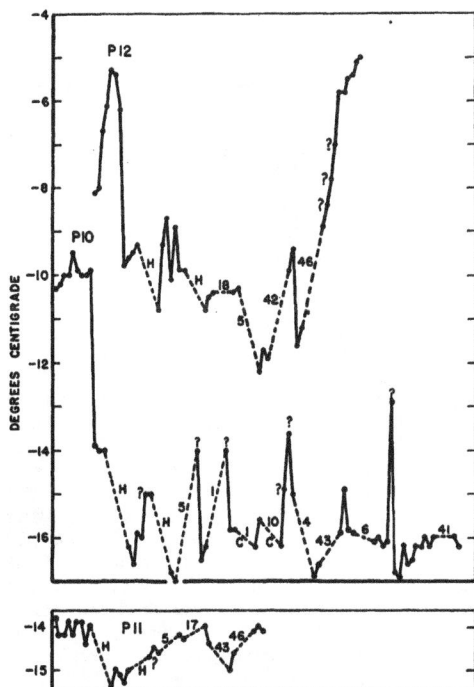

FIG. 25. Successive distillates and their residue: II. $P16$ and $P14$ vacuum distilled from $P15$ in that order, pump exhausted; $P15$ residue. Heatings (H) ·same for all, left to right: 2.2 and 4.2 hours. See section B,I,1,g.

by vacuum distillation from $P15$ (-7.9 to $-14.1°C$), to which they were successively sealed. In this case it is the second of the distillates ($P14$) that has a jagged graph and extends to the lowest temperature. The graph for the residue ($P15$) lies above each of the others, but does not differ much from that for $P16$.

From these confusing data it seems that the graph for the residue is generally less jagged than is that for the distillate, but need not lie far from it, and may lie appreciably below it. It is unfortunate that the reservoir was small for each of these P-bulbs.

$C37$ (-11.4 to $-13.6°C$), figure 16, was charged from the boiler used for steaming the bulbs. Presumably, it contains more motes than ordinary distilled water, and its concentration in mote-material is greater. It might be expected to have a graph somewhat like those for residues. Its graph is indeed less jagged than that for residue $P11$, and lies above it.

$P20$ (-6.8 to $-13.8°C$), figure 12, was charged from the same source as $C37$, but after the water had remained in a flask closed by a rubber stopper and undisturbed for about 2 years. The graph starts high, but quickly drops to near the early values for $C37$. The next freezing was 40 months later, and t_{sf} was again near that for $C37$. Six months later and in the dusty room, the bulb's inconveniently long neck was shortened, opening the bulb for a few minutes. Immediately thereafter t_{sf} was $-8.9°C$, and soon rose to

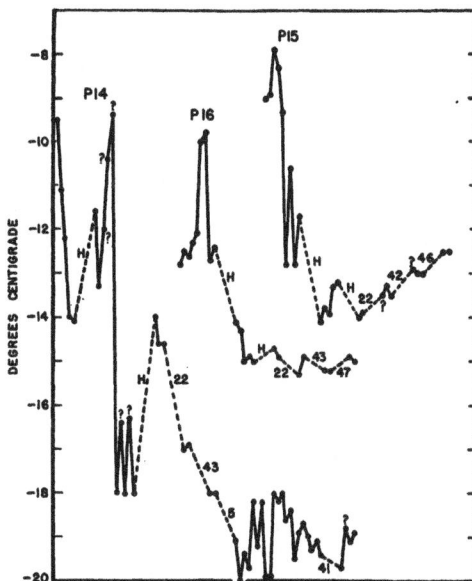

FIG. 24. Successive distillates and their residue: I. $P10$ and $P12$ vacuum distilled from $P11$ in that order, pump exhausted; $P11$ residue. Heatings (H) same for all, left to right: 2.2 and 4.2 hours. See section B,I,1,g.

− 6.8°C. It is thought that this change was caused by the entrance of air-borne motes. Compare this with C12 (fig. 15), opened in the cleaner room.

(7) *Washington City Water.*—C35 (− 6.0 to − 22.0°C), figure 14; C46 (− 4.4 to − 10.8°C), figure 26; C47 (− 4.6 to − 13.9°C), figure 10; C48 (− 5.9 to − 14.3°C), figure 3; and P8 (− 6.0 to − 14.1°C), figure 4, were each charged with water from the cold-water faucet in the laboratory. C35 was charged after much water had been drawn; C46, C47, and C48 were charged 10 days later, and differ only in that C46 and C47 contain much 2-mm copper wire, C48 does not; P8 was charged 2 weeks later than C46 to C48.

C33 (− 6.5 to − 14.0°C) and C34 (− 6.0 to − 12.9°C), figure 3, were charged from the hot-water faucet of the laboratory on the same day. C33 was charged first; no water had been previously drawn for a week or more, and that used was nearly at room temperature. Then water was drawn until it came from the faucet fairly hot (it was never very hot), and C34 was then charged. It will be noticed that the graphs for these differ but little from that, in the same figure, for C48, charged from the cold-water faucet. It is possible that the drain on that portion of the hot-water system was at that time so great that the water never became much heated; the subject was not further investigated.

(8) *Aquarium Water.*—C42 (− 7.2 to − 16.3°C), figure 12, was charged with water taken from an aquarium at a point below the surface and distant from the walls and the plants; C43 (− 7.4 to − 15.2°C), figure 11, with water from the same aquarium, but from points near the alga-covered walls and the aquatic plants.

(9) *Snow and ice.*—C15 (− 9.7 to − 15.8°C), figure 9; C23 (− 5.8 to − 10.0°C), figure 4; C24 (− 5.7 to − 14.7°C), figure 10; C32 (− 4.0 to − 5.6°C), figure 11; and P7 (− 7.2 to − 16.1°C), figure 15, were charged with melted snow. The charge in C32 was from snow-water that had been bottled a year earlier by another. The others were charged with the melt of newly fallen snow taken from just below the surface of a layer about 8 cm deep. C15 was charged from the midlayers of the melt; some of the melt was poured into C23 and C24, the latter receiving the smaller amount of surface scum; and P7, like C15, was charged from the midlayers, but a month later.

C45 (− 6.6 to − 12.0°C), figure 16, was charged with the melt of a clear selected crystal of manufactured ice. Its graph lies entirely above that for P11 containing the residue from a distillation, and also above that for C37, which was charged from the boiler used for steaming the bulbs.

(10) *D. C. Surface Waters.*—C38 (− 3.6 to − 11.9°C), figure 26; P4 (− 5.7 to − 9.7°C), figure 11; P5 (− 3.5 to − 11.3°C), figure 10; and P6 (− 5.6

to − 9.3°C), figure 11, were all charged with water from a clear swiftly flowing brook. C38 was charged directly from the brook; P4, P5, and P6 were charged from the midlayers of a volume of 9-day-old water taken from the brook at a point just below that at which drainage from a swamp entered it. Pieces of 2-mm copper wire were placed in P5, and to the brook water in P6 was added an equal volume of water from the cold-water faucet.

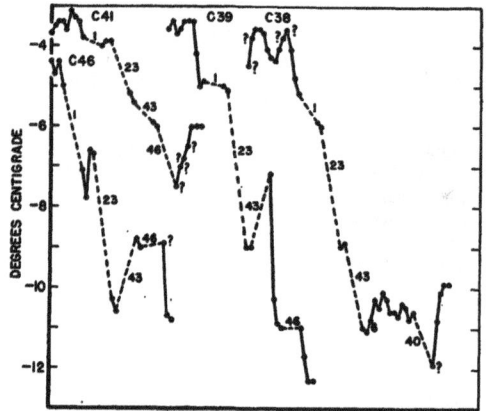

FIG. 26. Natural waters (D. C.). C38 clear brook; C39 stagnant pool, below surface; C41 bottom layers of same pool; C46 cold-water faucet, long pieces of Cu wire. For other D. C. waters, see references in text. See section B,I,1,g.

C39 (− 3.4 to − 12.3°C), figure 26; C40 (− 6.0 to − 11.3°C), figure 4; and C41 (− 3.1 to − 7.5°C), figure 26, were charged with water from a pool in marshy ground near the brook. The charge in C39 came from below the surface; that in C40, from the surface; and that in C41, from the bottom, including some of the bottom material. During the 9.5 years that these specimens have been under observation, not a single observed value of t_{sf} for C41 was nearly so low as were the low values for the other two.

(11) *Maryland Waters (Surface and Subterranean).* —The following bulbs were charged with waters obtained in Maryland, about 22 miles from Washington. C25 (− 5.1 to − 11.1°C), figure 4, was charged from a large, rapidly flowing spring; C26 (− 3.9 to − 8.4°C), figure 11, from a small, alga-covered, slow flowing one, adjacent to the other; C28 (− 4.5 to − 11.3°C), figure 10, from a swiftly flowing river into which those springs flow; C27 (− 3.9 to − 7.8°C), figure 3, from a pool in a marsh near the springs, bottom material is included. The water in C29 (− 6.6 to − 16.2°C), C30 (− 14.4 to − 16.2°C), and C31 (− 7.9 to − 17.3°C), figure 13, and P35 (− 12.7 to − 14.2°C), figure 21, came from a 60-ft well. C29 and C30 were charged with water from a tank supplied from the well, C29

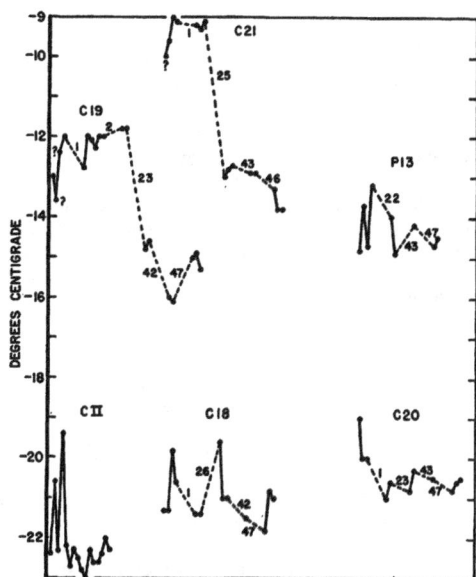

FIG. 27. Alcohol-water mixtures. *C*II vacuum distilled water into which ethyl alcohol from the bath leaked through a crack. At end of last plotted point (− 22.3°C), the neck cracked from the bulb; resealed, but imperfectly; then froze at − 25.3 and − 24.1, after which the concentration was about 14 per cent by weight. No more freezings. Seventeen days later, contents were decanted into *C*18, and *C*18 was sealed. *C*19 ordinary distilled water from that used for preparing solutions: *C*20 (20 per cent) and *C*21 (5.6 per cent). *P*13 ordinary distilled water containing an unknown amount of alcohol and an unperforated glass bead. See section B,I,1,*g*.

from the cold-water faucet, *C*30 from the hot; *C*31 was charged directly from the pump; and *P*35 was charged in the dusty room with water brought from the pump. Why the graph for *P*35 lies so little above that for *C*30 is not now known.

(12) *Mixtures of Water and Alcohol.*—Graphs for certain mixtures of water and ethyl alcohol are shown in figure 27. *C*II contained vacuum distilled water (air removed by boiling) into which alcohol from the chilling bath entered through a minute crack in the seal; later, after the last point (− 22.3°C) shown on the graph, the neck broke off; an attempt to make repairs was unsuccessful, and the contents were transferred to *C*18. The temperature at which the contents of *C*II was in equilibrium with ice, just before the transfer to *C*18, was found to be about − 6.5°C, corresponding to a concentration of about 14 per cent by weight. *C*19 is charged with a specimen of the water used in making the mixtures in *C*20 (19.8 per cent by weight) and in *C*21 (5.6 per cent by weight). *P*13 contains an unperforated glass bead and ordinary distilled water into

which an unknown amount of alcohol leaked through an imperfect seal; repairs were made before the first freezing.

By reference to table 7, it will be seen that no specimen, except certain surface waters, froze spontaneously at a temperature above − 4°C; that every natural water, including that from the Washington City water supply, froze spontaneously at or above − 9.7°C, whereas only 12 out of 40 specimens of distilled water did so. On the other hand, only 2 of the 40 distilled specimens failed to supercool beyond − 13.5°C; whereas only 11 out of 27 specimens of natural waters

TABLE 7

EXTREME OBSERVED VALUES OF t_{sf} FOR EACH SPECIMEN

In each class, the specimens are arranged in order of falling temperatures—first for the highest observed t_{sf}; second for the lowest. The data for distilled well-water and for mixtures of alcohol and water are not given here.

Highest and lowest observed values of t_{sf} (°C)
for each specimen

Surface waters: D.C.				Ordinary distilled water: Stock			
C41	− 3.1 − 7.5	C41	− 3.1 − 7.5	C12	− 6.3 −15.9	C49	− 8.4 −13.6
C39	− 3.4 −12.3	P6	− 5.6 − 9.3	C13	− 7.2 −13.9	P30*	− 9.8 −13.7
P5	− 3.5 −11.3	P4	− 5.7 − 9.7	C49	− 8.4 −13.6	C9	−11.4 −13.8
C38	− 3.6 −11.9	C40	− 6.0 −11.3	P30*	− 9.8 −13.7	C13	− 7.2 −13.9
P6	− 5.6 − 9.3	P5	− 3.5 −11.3	C8	−10.6 −14.4	C8	−10.6 −14.4
P4	− 5.7 − 9.7	C38	− 3.6 −11.9	C9	−11.4 −13.8	C12	− 6.3 −15.9
C40	− 6.0 −11.3	C39	− 3.4 −12.3	C19	−11.8 −16.1	C17	−12.9 −16.0
				C17	−12.9 −16.0	C19	−11.8 −16.1
Surface waters: Md.				**Ordinary distilled water: Special**			
C26	− 3.9 − 8.4	C27	− 3.9 − 7.8				
C27	− 3.9 − 7.8	C26	− 3.9 − 8.4	P23	− 8.3 −14.1	P23	− 8.3 −14.1
C28	− 4.5 −11.3	C25	− 5.1 −11.1	P25	− 9.5 −17.1	P22	−10.5 −15.7
C25	− 5.1 −11.1	C28	− 4.5 −11.3	P22	−10.5 −15.7	P24	−11.4 −15.8
				P27	−10.9 −16.6	P26	−12.0 −16.7
Subterranean: Md.				P24	−11.4 −15.8	P25	− 9.5 −17.1
C29	− 6.6 −16.2	C29	− 6.6 −16.2	P26	−12.0 −16.7	P27	−10.9 −16.6
C31	− 7.9 −17.3	C30*	−14.4 −16.2				
C30*	−14.4 −16.2	C31	− 7.9 −17.3	**Special: Quartz still**			
				G3	−10.6 −16.9	G2	−11.0 −16.3
Washington City water				G4	−10.9 −17.5	G3	−10.6 −16.9
C46	− 4.4 −10.8	C46	− 4.4 −10.8	G2	−11.0 −16.3	G5	−11.6 −17.1
C47	− 4.6 −13.9	C34	− 6.0 −12.9	G1	−11.2 −17.3	G1	−11.2 −17.3
C48	− 5.9 −14.3	C47	− 4.6 −13.9	G5	−11.6 −17.1	G4	−10.9 −17.5
C35	− 6.0 −22.0	C33	− 6.5 −14.0				
P8	− 6.0 −14.1	P8	− 6.0 −14.1	**Conductivity water**			
C34	− 6.0 −12.9	C48	− 5.9 −14.3	C10	−10.0 −16.0	P2	−10.3 −15.7
C33	− 6.5 −14.0	C35	− 6.0 −22.0	G2	−11.0 −16.3	C11	−11.8 −15.7
				P1	−10.6 −16.3	C10	−10.0 −16.0
Snow				C11	−11.8 −15.7	P1	−10.6 −16.3
C32^b	− 4.0 − 5.6	C32^b	− 4.0 − 5.6				
C24	− 5.7 −14.7	C23	− 5.8 −10.0	**Vacuum distilled: Boiled out**			
C23	− 5.8 −10.0	C24	− 5.7 −14.7	P29*	− 7.6 −11.3	P29*	− 7.6 −11.3
P7	− 7.2 −16.1	C15	− 9.7 −15.8	P31*	− 7.8 −14.0	P31*	− 7.8 −14.0
C15	− 9.7 −15.8	P7	− 7.2 −16.1	C14	− 9.8 −21.1	P21*	−12.1 −15.8
				CI	−11.3 −20.4	CIII	−15.1 −19.1
Ice				P21*	−12.1 −15.8	CIV	−12.2 −20.1
C45	− 6.6 −12.0	C45	− 6.6 −12.0	CIV	−12.2 −20.1	CI	−11.3 −20.4
				CIII	−15.0 −19.1	C14	− 9.8 −21.1
Aquarium							
C42	− 7.2 −16.3	C43	− 7.4 −15.2	**Vacuum distilled: Pumped out**			
C43	− 7.4 −15.2	C42	− 7.2 −16.3	P12	− 5.0 −12.2	P12	− 5.0 −12.2
				P14	− 9.4 −20.0	P17	− 9.8 −13.7
Residues				P10	− 9.5 −17.0	P16	− 9.8 −15.0
P15	− 7.9 −14.1	P15	− 7.9 −14.1	P16	− 9.8 −15.3	P10	− 9.5 −17.0
P18	−10.6 −15.7	P11	−13.8 −15.4	P17	− 9.8 −13.7	P14	− 9.4 −20.3
C7	−12.7 −16.3	P18	−10.6 −15.7				
P11	−13.8 −15.4	C7	−12.7 −16.3	**Distilled at 10 to 15 cm-Hg**			
				P32*	−13.0 −16.0	P32*	−13.0 −16.0
Boiler							
P20	− 6.8 −13.8	C37	−11.4 −13.6	**Other distilled waters**			
C37	−11.4 −13.6	P20	− 6.8 −13.8	P34*	− 9.0 −18.1	P33*	−12.0 −14.6
				P28*	− 9.5 −15.0	C50*	−13.6 −14.7
				P33*	−12.0 −14.6	P28*	− 9.5 −15.0
				C50*	−13.6 −14.7	P34*	− 9.0 −18.1

* *C*30 contains preheated water.

^b *C*32 was charged from a bottle of melted snow collected a year previously by another than the author.

* *C*50, *P*21, and *P*28 to *P*34, inclusive, were prepared in the dusty room, but the charging of *C*50 and of *P*28 was done in a nominally dust-free box.

(including the City supply) ever supercooled that far. Nevertheless, excluding mixtures with alcohol, the greatest supercooling observed in this study (22.0°C) was for a specimen (C35) of Washington City water, which early in its study froze at − 6.0°C.

Whence it seems that the condition responsible for spontaneous freezing at the higher temperatures is generally present in natural waters, but is destroyed by heating, by aging, and by the process of distillation; and that although distillation favors supercooling, the condition for great supercooling may arise without it.

l. Main Conclusions

The study of the sealed specimens has led to the following conclusions, which must be satisfactorily explained by any acceptable theory of the phenomenon of spontaneous freezing.

1. Freezing is determined by the temperature of the specimen, not by the duration of its chilling.

2. The time distribution of the observed temperatures at which a given specimen freezes spontaneously is not chaotic.

3. For an unchanging specimen, t_{sf} remains constant.

4. For an actual specimen, t_{sf} may change monotonously with the time. The change may be in either direction. When t_{sf} is initially high, the change is usually a fall in temperature; seldom has a monotonous rise been observed until after t_{sf} had fallen to a low temperature.

5. For each specimen there are one or more preferred temperatures at which spontaneous freezing occurs.

6. A particular preferred temperature may be a characteristic of only a single specimen; on the other hand, it may be common to several specimens.

7. A particular preferred temperature may be approached either from above or from below; and for any given specimen, the direction of approach may vary from time to time.

8. If a specimen has several preferred temperatures, the observed values of its t_{sf} may vary from one to another in any order, even ignoring intermediate ones. Such variations may be so frequent and rapid as to give a superficial impression of chaos.

9. Such rapid changes in t_{sf} seem to characterize specimens that contain relatively few effective motes, of various kinds.

10. Prolonged chilling has no significant effect upon the value of a specimen's t_{sf}.

11. Preheating the melt produces no certain effect upon its t_{sf} if that, prior to the heating, lay at or below a preferred temperature within the range − 11 to − 15°C.

12. If prior to the heating, the specimen's t_{sf} lay at a higher temperature than that just indicated, then the heating reduced its t_{sf} to such a preferred temperature.

13. When t_{sf} is so lowered by the heating, that lowering is permanent; i.e., t_{sf} is not raised by subsequent freezings.

14. In general, fresh specimens of natural waters freeze spontaneously at higher temperatures than do distilled waters.

15. The condition responsible for that initial high temperature of freezing is destroyed by age, by heat, and by the process of distillation.

16. As a result of those processes, a specimen of natural water may presently attain a condition in which it can be greatly supercooled (− 22°C), even more than has been observed for any distilled specimen studied in this work.

m. Mixtures

How is the spontaneous-freezing-point of a mixture of two specimens of water related to those of its constituents? Many potential sources of error and uncertainty attend an experimental investigation of that very important question.

In its essence, such an investigation consists in the determination of the spontaneous-freezing-point of a specimen of each of the constituents, each contained in a closed bulb (or other container) of its own, each bulb being capable of being opened and reclosed as may be desired; then each bulb is opened, a portion of the contents of one is transferred to the other, the bulbs are reclosed, and the temperature of spontaneous freezing is determined for the contents of each. Then there may be another transfer; and so on. The transfer will preferably be made by means of a pipette or dropper; so that the composition of the mixture at each step may be known.

Or a single bulb that can be opened and closed may be used, the water that is to be added from time to time being taken from the midlayers of a relatively large volume. Then a specimen of that water is placed in the bulb and frozen, and that observed temperature of spontaneous freezing is assumed to be typical of all such specimens that may be taken. The bulb is then cleaned, charged with the other constituent, and the contents frozen. It is then opened, a known volume of the first water is added, the bulb is closed, and the contents frozen; and so on.

A weakness of the second method is the necessary assumption that each and every small volume of the first water—that added in the making of the mixture—freezes spontaneously at the same temperature as did the specimen that was taken as typical. But even in the first method there is a corresponding weakness, viz. the assumption that each small volume withdrawn from the bulb would freeze spontaneously at the same temperature as did the entire contents.

That is, each assumes that the smallest volume added in making the mixture is typical of the entire volume from which it was taken. As will be shown in a later section that is not in general true.

Thus may arise a fundamental uncertainty in addition to the ordinary experimental errors in determining the temperature of spontaneous freezing. To that is to be added the effect of the irregularities that are frequently observed in the case of new specimens—every time the contents of a bulb is added to or reduced, it becomes a new specimen. Were each new specimen allowed to stand until it had become stabilized by age, the test would be unduly prolonged and other difficulties might arise. There is also a danger that the motes responsible for the higher freezing of one specimen may be gradually dissolved as the amount of the lower-freezing specimen in the mixture is increased; and a danger of casual contamination every time a bulb is opened; and a danger of removing significant motes by their adherence to the pipette. The last is especially great if those motes tend to congregate in the free surface. That such congregation may exist, is suggested by the fact, already remarked upon, that there are indications that the contents of some of the sealed bulbs occasionally begin to freeze at the meniscus.

At the time that the study of mixtures began, the effect of motes in the water was only dimly recognized, and I was not at all sure that they played a prime role in the phenomenon under study. I did not properly appreciate the sources of error and uncertainty just mentioned. But as the study progressed, very annoying irregularities, later attributed to those effects, arose. As the more annoying of them can be largely avoided by always adding portions of the lower-freezing component to the whole of the other, and never conversely, little more need be said of them.

It had been suggested that the temperature to which a specimen of water can be supercooled—that is, its spontaneous-freezing-point—is determined by the concentration of substances dissolved in it, and especially by the concentration of hydrogen ions. Were that true, then that temperature for the mixture would vary continuously with its composition, and might be expected to lie, in general, between those for its constituents. And it would be a matter of complete indifference which specimen was gradually added to the other.

When the work was begun, results of that general kind were rather expected. But those obtained were quite different.

For example, to − 5.7°C water [s] was added − 9.8°C water, and the mixture froze at − 6.0°C; a week later it froze at − 6.2°C. When − 13.8°C water was added to − 6.0°C water, the mixture froze at − 6.0°C. When to − 8.0°C water were added 1, 2, 4, and 6 times its volume of − 13.6°C water, the mixtures froze at − 8.3, − 8.3, − 8.4, and − 8.0°C, respectively; when to − 6.6°C water were added 1, 2, 4, and 6 times its volume of − 13.7°C water, the mixtures froze at − 6.7, − 6.8, − 7.0, and − 6.8°C, respectively. And

a mixture formed by adding 0.6 ml of − 12.0°C water to 3 ml of − 9.0°C water froze at − 9.0°C.

The mixture freezes spontaneously at the same temperature as does its higher-freezing component, whatever the relative proportions of its two components.

But when the reverse procedure was followed—when portions of the higher-freezing component were added to the whole of the lower-freezing one—the results were very erratic. Sometimes the mixture froze at the same temperature as did its higher-freezing component, exactly as in the cases just reported. At other times, at some intermediate temperature, but there was no obvious relation between that temperature and the composition of the mixture, beyond the fact that the mixtures that froze at the intermediate temperatures seldom contained more than a small amount of the higher-freezing component.

A single example, in duplicate, will suffice. In each of two capped bulbs was placed 1.5 ml of water freezing at − 13.6°C, and to each were made 6 successive additions of 0.3 ml each, from a specimen freezing at − 6.0 to − 6.8°C. One set of 6 mixtures froze at − 11.0, − 7.2, − 7.0, − 6.9, − 6.9, and − 7.0°C, respectively; the other at − 12.0, − 7.8, − 7.9, − 7.8, − 7.0, and − 7.0°C, respectively.

None of these results accord with what would be expected if the temperature of spontaneous freezing were determined primarily by the concentration of dissolved substances, whether of hydrogen ions or of something else. But they do accord with what one would expect if that temperature were determined by the most effective mote contained in the specimen.

If the temperature were so determined, then one should expect intermediate values when portions of the high-freezing component are added to the lower-freezing one; because the portion transferred may contain none of the more effective motes, and it is the most effective mote that is transferred that determines the temperature at which the mixture freezes. On the contrary, when portions of the lower-freezing component are added to the higher-freezing one, the failure to transfer motes, of any kind, is of no importance, because the most effective mote of all, already in the higher-freezing component, is sure to remain in the mixture, unless the pipette by which the addition is made is dipped into the mixture. If that is done, the controlling mote, or motes, may adhere to the pipette and be removed when it is withdrawn. A few cases of that kind seem to have been observed.

Whence one may conclude that the temperature at which a binary mixture of waters freezes is that which is characteristic of the higher-freezing of the two components, it being assumed that neither component is modified in any way in the process of making the mixture.

Dehlinger and Wertz [25] have more recently reported a similar result for the crystallization of supersaturated solutions.

[s] For brevity "− t°C water" is used to denote water that freezes spontaneously at − t°C.

n. Effect of Solutes on Thermal Initiation of Freezing

How does the presence of dissolved substances affect the temperature at which a specimen of water freezes spontaneously?

Reasons have already been given for believing that dissolved glass produces no effect. The obtaining of a more direct experimental answer to this very important question is not easy, involving as it does the ever-present danger of casual contamination, and of decontamination if the number of effective motes is small.

Two procedures have been followed in seeking the answer. The first was a qualitative one, in which successive amounts of a solution of unknown concentration were added to a specimen of water contained in a test tube closed by a paraffined cork through which passed a thermometer dipping in the water. The water was frozen in the manner already described for the sealed bulbs, but its temperature was taken as that of the thermometer dipping in it. Those observations indicated that any effect produced on t_{sf} by the presence in solution of small or moderate amounts of HCl, NaCl, NaNO$_3$, or sucrose, is exceedingly small—not over a degree at the most.

The second procedure was a quantitative one, and capped bulbs were used. A solution of known concentration of the substance to be used was made up with distilled water from the same source as that to be studied. Into one of two capped bulbs in which the water under study froze at the same temperature was placed a charge of the solution; and into the other was placed a charge of the water to be studied. The temperature at which each froze spontaneously was determined. Then by means of a small rubber-bulb pipette, successive small known amounts of the solution were transferred from the bulb containing it to that containing the water, or the reverse, and the several resulting values of t_{sf} were determined. From time to time the t_{sf} of the solution, or water, remaining in its bulb was also determined. Whether the observations should proceed in the direction of increasing concentration, or the reverse, depends upon the spontaneous-freezing-points of the initial solution and the water. As has been shown in the preceding section, the liquid freezing at the lower temperature should be added to the other. Failure to realize the importance of this precaution led to much unnecessary and useless work. Typical sets of data are given in table 8.

TABLE 8

EFFECT OF SOLUTES ON SPONTANEOUS FREEZING

The numbers tabulated are the observed spontaneous-freezing-points of the several specimens; A, C, D, F, and H are the designations of the 5 capped bulbs that were used.

In the first series, successive small amounts of a molal NaCl solution were transferred from C to A; the observed t_{sf} of the resultant solution in A fell irregularly from −13.3°C for water to −14.9°C for 0.4M NaCl, corresponding to a depression of 4°C

for a molal solution. But, at the same time the t_{sf} of the continually decreasing volume of the molal solution in C fell by the same amount. The cause of that fall is not clear; it may have resulted from a removal of high-freezing motes by, or on, the pipette, but more likely from a progressive decrease in the size of the motes.

In the next series, successive small amounts of water were transferred from F to the sucrose solution in D. The observed t_{sf} of the resulting solution in D rose fairly monotonously from −10.0°C for the molal sucrose solution to −8.0°C for the 0.3M one, corresponding to a depression of 3°C for a molal solution. Then a part of the 0.3M solution was transferred to H, and more water was added to F. The resulting lowering of the t_{sf} of each indicates that high freezing motes had been removed by the pipette at some time, or had decreased in size. Continued dilution resulted in an additional rise of about 0.5°C in t_{sf} for a decrease in concentration of 0.15M, again about 3°C for a molal solution.

A Solution (NaCl) Concen.	A t_{sf}	C 1M NaCl t_{sf}	D Solution (Sucrose) Concen.	D t_{sf}	F Water t_{sf}	H 0.3M Sucrose t_{sf}
Zero	−13.4	−13.3	1M	−10.0	−11.5	—
Zero	−13.2	−13.0	1M	− 9.9	−11.3	—
0.025M	—	—	0.9M	—	—	—
0.025M	−13.8	—	0.9M	− 9.7	—	—
0.025M	−13.6	—	0.9M	− 9.5	—	—
0.05M	—	—	0.8M	—	—	—
0.05M	−13.2	—	0.8M	− 8.8	—	—
0.05M	−13.2	—	0.8M	− 9.0	—	—
0.1M	—	—	0.7M	—	—	—
0.1M	−13.7	—	0.7M	− 8.8	—	—
0.1M	−13.7	—	0.7M	− 8.6	—	—
0.2M	—	—	0.6M	—	—	—
0.2M	−13.9	—	0.6M	− 8.0	−11.4	—
0.2M	−13.9	—	0.6M	− 8.3	—	—
0.4M	—	—	0.6M	− 8.6	—	—
0.4M	−14.0	−14.7	0.6M	− 8.7	—	—
0.4M	−14.2	−14.7	0.5M	—	—	—
0.4M	−14.7	—	0.5M	− 8.4	—	—
0.4M	−14.6	—	0.5M	− 8.3	—	—
0.4M	−14.8	−15.0	0.4M	—	—	—
0.4M	−14.9	−15.0	0.4M	− 8.1	—	—
0.4M	—	—	0.4M	− 8.1	—	—
0.4M	−14.9[e]	−14.9[e]	0.3M	—	—	—
			0.3M	− 8.1	—	—
			0.3M	− 7.9	—	—
			0.3M	− 9.4[a]	−13.0[b]	−7.9[c]
			0.3M	− 9.3	−12.8	—
			0.25M	—	—	—
			0.25M	− 9.0	—	—
			0.25M	− 9.0	—	—
			0.2M	—	—	—
			0.2M	− 9.0	—	—
			0.2M	− 8.9	—	—
			0.15M	—	—	—
			0.15M	− 8.9	—	—
			0.15M	− 8.8	—	—
			0.1M	− 8.9	−12.6	−8.0
			0.1M	—	−12.9	—
			0.1M	− 9.9[d]	—	−7.8[d]
			0.1M	− 9.9	—	−7.9

[a] Remainder of solution after placing a portion in H.
[b] More water has been added to F.
[c] A portion of solution from D.
[d] Observations 1 week later than the preceding.
[e] Observations 5 days later than the preceding.

In spite of irregularities and of an occasional progressive change in the casual contamination, the observations show that the presence of completely dissolved solute that does not act chemically upon the motes certainly did not lower t_{sf} by more than the amount by which it would have depressed the normal melting point (less than 4°C for a normal solution of NaCl). Later it was found that Füchtbauer [36] had reported that soluble additions to organic substances produced a very small effect, of the order of their freezing-point depressions, scarcely exceeding his experimental error.

Consequently, the effect of the dissolved substances contained in the specimens of water that have been studied, even in the natural waters, is entirely negligible as compared with the observed variations of t_{sf} from specimen to specimen, and with the changes that attend prolonged heating and the aging of a specimen.

o. Effects of Changes in the Motes

It was assumed in the paper of 1938, and in some of the discussion on the preceding pages, that the t_{sf} of a specimen containing motes of a single substance is determined by the size of the most efficient (spherical) mote present. Since the theory, to be discussed later, assumes that it is the curvature of the interface that determines the efficacy of any region of a mote of a specified substance, a nonspherical mote may be regarded as an aggregation of spherical ones, each acting independently of the others. Hence only spherical motes need be considered; and by "size" is meant the size of the equivalent spherical mote.

In the earlier report it was stated that the most efficient mote of a given substance is the largest that does not exceed a certain size. It is obvious, however, that such a statement can refer only to motes that are smaller than that particular size. The more general statement is this: For a given melt, the most efficient mote of a given substance is that which most nearly approaches a certain size.

No mote of that substance can initiate freezing in that melt at a temperature above the t_{sf} corresponding to that particular size. That t_{sf} is a unique, preferred temperature. As the size of a mote decreases, its efficacy (the temperature at which it can initiate freezing) rises if the mote is larger than that particular size, and falls if it is smaller. The former may be called a "major" mote; the latter a "minor" one. In general, a specimen has as many preferred temperatures as there are mote-materials in the melt; although none can be observed in the presence of motes that can initiate freezing at a higher temperature.

A detailed discussion of the theory may be found in the portion of this paper devoted to the theories of the initiation of freezing. Here it will suffice to point out certain consequences, and to remark that the importance of the larger motes and the significance of preferred

temperatures were not at all clearly perceived until the very end of the experimental work.

From what has gone before, it follows that as a result of a continued solution of the motes t_{sf} may either rise or fall, remain essentially constant, cluster about certain preferred values, change in either direction either slowly or abruptly; in brief, t_{sf} may vary in any of the ways that have been observed in this work.

Similar changes may arise from a continued growth of a mote by deposition from solution or by coalescence of two or more motes, or from a dispersal of coalesced motes.

The possibilities of the theory are ample to account for all the observations reported in the preceding text. One may, however, validly ask for experimental evidence that such changes in the motes do indeed give rise to those changes in t_{sf}. Since the motes are far too small to be seen and handled individually, it is necessary to rely on inference. Even so, a presumptively good case can be made.

Heating and aging will presumptively lead to a decrease in the size of the motes; and if the controlling mote is a minor one, to a fall in t_{sf}. Such a fall is observed if the specimen initially froze at a high temperature.

The very active chromic cleaning solution (potassium bichromate and sulphuric acid) may be expected to dissolve many kinds of mote material. To a 3-ml specimen of faucet water freezing at -7.0°C was added 0.1 ml of the cleaning solution. After a half hour, $t_{sf} = -9.9$°C; after 19.3 hours, -15.0°C; after 43.3 hours, -14.9°C; and after 143 hours, -14.9°C. During the first 19 hours, t_{sf} fell 8.0°C, and remained thereafter unchanged for at least 5 days. This test was made in a capped bulb, and indicates that a reduction in the size of the mote is accompanied by a fall in t_{sf}.

In a later experiment made in the dusty room and with the water in a large test tube closed with a paraffined cork, the addition of the cleaning solution to distilled water freezing at -6.8°C (probably as a result of contamination by air-borne motes) produced no certain effect. Perhaps those particular high-freezing motes were not readily soluble in the solution.

If the effective motes are not too small, they should be removable by filtration. For two capped bulbs (C and D) charged from the same lot of water, t_{sf} was -6.8 and -7.0°C, respectively. The bulbs were then emptied, washed well with the filtrate obtained by passing water from the same lot through a folded filter of Whatman's No. 2 paper, and were charged with the filtrate. For those charges, t_{sf} was -11.9 and -12.0°C, respectively. Filtering had lowered t_{sf} by 5°C, presumably by the removal of the larger motes.

It should be possible to raise the value of t_{sf} by adding more efficient motes. To water freezing at -12.3°C were added fibers of Baker's "washed and ignited" asbestos that had been washed in water from the same source, and had been ignited just before they were placed in the water; t_{sf} rose at once to $-6.9°C$, and was not changed by the addition of more asbestos. The addition of bits of new, unwashed, black rubber tubing with its attendant bloom raised t_{sf} from $-13.0°C$ to $-9.5°C$, where it remained unchanged for 3 weeks. Addition of cork dust to $-13°C$ water raised t_{sf} to $-12°C$, and a further addition of filings of ordinary solder raised it to $-11°C$. The ash from burnt sealing wax raised t_{sf} from -9 to $-5°C$; cigarette ashes, from -9.5 to $-4.0°C$, and on standing over night to $-2.7°C$, but during successive freezings t_{sf} fell again to $-5°C$.

That a single mote may seize control from others that are effective only at lower temperatures, is shown by the following experiment. Into capped bulb G containing water freezing at $-13.6°C$ was placed a bit of cinder; t_{sf} rose at once to $-5.8°C$, 16 months later to $-5.2°C$, and after 5 years to $-5.0°C$. The water from G was then transferred to bulb E in which a specimen of a certain water had just frozen at $-12.1°C$; and the transferred water froze at $-12.3°C$. Water from the same source was placed in G, containing the cinder and wet with the old water, and it froze at $-4.9°C$. Then G was emptied, the cinder removed, and G was well washed and charged with the same $-12°C$ water. That charge froze at $-12.0°C$. There can be no doubt that the cinder was responsible for the freezings around $-5°C$; the freezings always started at the same point on the cinder.

Since for natural waters exposed to the air t_{sf} is never many degrees below $0°C$, it may be inferred that air usually carries high-freezing motes, and that the casual contamination of a lower-freezing specimen by exposure to the air will usually result in the addition of such motes, and a consequent rise in t_{sf}. Since the water will be far from saturated with the material of such newly added motes, there may be marked changes in the sizes of those motes during a short time thereafter, and corresponding changes in t_{sf}. The change in t_{sf} may be in either direction, and need not be monotonous. Such may be, but is not necessarily, the explanation of the irregularities in t_{sf} that are frequently observed early in the life of a sealed specimen and soon after a capped bulb is closed.

2. PROGRESSIVE FREEZING

Usually, the spontaneous freezing of supercooled water contained in a closed bulb seems to the eye to occur instantly throughout the volume, to be a cataclysmic event. Were one's observations quite limited, one might infer that the phenomenon of spontaneous freezing is essentially cataclysmic, and that a different mechanism is involved in those rare cases in which there is little or no evidence of a cataclysm.

Such inference would, however, not be justified. An apparent cataclysm will occur if the melt contains a very large number of effective motes of the same size, scattered throughout its volume. Freezing will begin at each of those motes at the same time, giving one the impression of a cataclysm. If those effective motes were confined in large part to a restricted region in the melt, freezing would seem to begin there, and to sweep out from that region. If there were but one effective mote, freezing would begin there, and grow out from that point, the water changing to ice only at its boundary with the advancing ice and on account of that ice; the water would not actively freeze, but would be passively frozen by the ice that touches it. All these cases have been observed.

During the study of the sealed bulbs, the observer now and again, at rare intervals, obtained the impression that a wave of freezing had passed through the liquid. Normal solutions of NaCl froze progressively and rapidly, but not from a single clearly defined point. Sometimes they seemed to freeze instantly across the entire meniscus, and a wave of freezing, extending entirely across the bulb, swept downward. Solutions of $NaNO_3$ also froze progressively, but from the bottom upward. Solutions of sucrose froze quite slowly, usually from the top downward, and from a single point if the solution was not too dilute. Mixtures of alcohol and water always froze from a single point, and the ice grew slowly; ice first appeared as a small rosette, sometimes on the bottom of the bulb, sometimes in the meniscus, sometimes on the side of the bulb, and the ice grew slowly from the rosette until the entire volume of the liquid had become solidified, about 25 seconds being required in the case of CII, containing about 14 per cent of alcohol by weight. In one case, the rosette formed at or near the bottom of the bulb, broke loose, rose to the meniscus, and continued its growth there. The freezing of such mixtures has never, in this work, been observed to proceed from more than a single point; for this, I have no explanation to offer. In the case of the cinder, considered in the preceding section, freezing always started at the same point, and the ice grew progressively, but rapidly, from that point.

3. APPEARANCE OF THE ICE

a. Volume

Numerous notes concerning the appearance of the ice in the bulbs were recorded from time to time, but they do not seem to lead to any general conclusion and are not satisfactorily amenable to a summarizing. In some cases there were clear evidences of planes, sometimes intersecting along a line; in other cases there appeared to be radiating needles. Generally, the bulb was removed as soon as the water froze, and the ice was cloudy throughout, with little, if any, indication of a distinctive structure.

An interesting observation may be mentioned here. The frozen mass always contains unfrozen water, and consequently is denser than ice, and floats less high.

After melting has proceeded so far that the surface has become smooth, the surface rapidly acquires a curvature similar to that of a liquid meniscus, and there is little to indicate the boundary between the wet surface of the ice and the surface of the surrounding annulus of water. If after sufficient melting the ice be turned bottom up, the convex bottom will protrude well above the surface of the water, but will rapidly melt, take on the meniscus-like curvature, and float as before, but with its former bottom up. If turned on its side, the same sequence of events is observed.

The rapid melting of the projecting portions can reasonably be explained by the condensation of vapor distilled from the adjacent liquid, the vapor pressure over ice being less than that over water at the same temperature. But to me it seems surprising that the density of the ice-and-water mass, combined with the weighting down of the mass by the water carried in its concave surface and with the capillary pull of the water surface between the ice and the wall of the bulb, should be such as to make the mass float with its surface so nearly in the surface of the water. These observations have been strictly casual, and have had to do with bulbs from which the air had been removed by prolonged boiling, but in all these cases the frozen mass, after adjustment by melting, floated so low that by means of the unaided eye it was very difficult to determine whether the ice did indeed project above the surface of the water.

b. Surface

The surface of the frozen water is nearly always rough, being marked by hillocks and spurs and spicules. The last are particularly prominent when the bulb contains very little air. Spicules often grow with surprising speed; frequently a spicule will shoot out from the surface to a length of a centimeter or more in the twinkling of an eye. But sometimes they grow much more slowly. They have been described by several previous observers [5, 8, 27a, 31, 58].

Some have thought that they grow by reverse sublimation, by the capturing of vapor molecules by the ice. But that explanation, especially in the case of rapid growth, seems to me to be quite unsatisfactory. For one thing, the growth of a star of ice, to be presently described, was exceedingly slow (about 6 mm in 10 minutes), and that was by reverse sublimation when there was a large surface of unfrozen water to furnish the vapor. For another thing, the longer spicules observed in this work have appeared after the surface of the water had become coated with ice; and the same may have been true of the shorter ones. But after such coating, one has available for building the spicules only the excess of the amount of vapor contained in the vapor phase when in equilibrium with water, over that when in equilibrium with ice at the same temperature. If the cooling bath could remove the latent heat as

rapidly as it was liberated, keeping the temperature constant, at say $-10°C$, then there would be available for building the spicule only 0.22 micrograms [4] of vapor in each cubic centimeter of the gas phase, which when frozen will have a volume of only 0.24×10^{-6} cm³. Hence the longest spicule that can be so formed from 10 cm³ of the vapor, an amount exceeding that in any of the sealed bulbs, would correspond to a cylinder 0.1 mm in diameter and 0.3 mm long; whereas spicules several tenths of a millimeter in diameter and 10 mm long were not uncommon. The formation of such large spicules by reverse sublimation would seem to be impossible under those conditions.

But another possibility exists, as was pointed out in the paper of 1938, and as I have now found had been suggested earlier by Bally [8]. When a specimen that is supercooled freezes, the water very promptly becomes sealed up by a layer of surface ice that adheres to the wall of the bulb; and the temperature of the water rises to 0°C. As freezing proceeds, the pressure of the enclosed water rapidly increases, the temperature changing but little. The pressure may rupture the surface ice at some weak spot. A jet of very slightly supercooled water then issues through the break, its surface freezes promptly, forming a tube which grows at its tip, and through which water continues to flow until the pressure is sufficiently relieved or the tube has become blocked with ice. The growth is necessarily rapid and short-lived; and as the tube becomes blocked, its side wall may become ruptured, giving rise to one or more side branches. All true spicules observed during this investigation are believed to have been so formed. Which is not to say that spicules are never formed by reverse sublimation; but surely those so formed must grow very slowly and can be of more than diminutive size only under special conditions.

Weak spots where a break may occur, giving rise to spicules, may exist not only in the surface ice itself, but also at the contact of that ice with the wall of the bulb, and also in the thin layer of ice covering the wall of the bulb in the gas-phase, provided that between that layer and the wall and extending from the weak spot to the main volume of the liquid there is a continuous space, or tube, through which water can flow. Spicules that grew rapidly have been observed in such position. In 1939 v.Wartenberg [91] reported that when a cold solid is plunged into a warmer liquid the latter may not wet the solid and a layer of unfrozen liquid may, under some conditions, remain for a short time between the solid and the surrounding film of frozen liquid. Whether this has any bearing on the observations herein reported is not known.

---

[4] Computed as though the vapor were an ideal gas of molecular weight 18. In the analogous computation in my book [29: 638] there are two errors; as a consequence, the volume of vapor required to produce a spicule of the size there assumed is 16 cm³, not 45 cm³ as there stated.

That spicules occur more frequently in bulbs contain- ing little air than in those containing air at atmospheric pressure is perhaps to be explained by the greater cool- ing of the water surface in the former case, as a result of the more rapid distillation to the cooler wall of the bulb in the gas phase. That cooling will facilitate a prompt formation of surface ice.

Why some long spicules, formed after the surface of the water had become frozen, grew relatively slowly, taking perhaps a few seconds to grow a centimeter in length, is not clear. Perhaps the conditions were such that the water was freezing slowly both in the bulb and in the hypothetical tube of the spicule.

4. ICE ON WALLS OF BULBS

a. Below the Meniscus

When a specimen is plunged into a bath at a tem- perature that is lower than its t_{sf}, the water immediately adjacent to the walls may become cooled to its t_{sf} be- fore the rest of the liquid shall have become super- cooled. In that case, ice will form first at the walls, and will gradually thicken, the innermost volume of the water never falling below 0°C until solidification is complete. In the early stages of growth, this trans- parent layer of ice on the wall is hard to detect by the unaided eye, but it very soon becomes thick enough to be plainly visible when viewed between crossed Polar- oids. When the entire volume of water is allowed to solidify under such conditions, the resulting ice is fre- quently very clear, except in the center, where there are many bubbles of entrapped air.

Observations reported by v.Wartenberg [91] sug- gest that there may occasionally exist for a very short interval a thin unfrozen film between the wall and the ice.

b. Above the Meniscus

Records concerning the formation of ice and of frost upon the portion of the wall of the bulb that is in con- tact with the gas-phase are very deficient.

(1) *Frost.*—I have been impressed by the general absence of frost on the walls before the bulk of the liquid has frozen. At the instant of freezing, the wall in the gas phase frequently, perhaps one may say gen- erally, becomes coated with a filigree of ice. But there are exceptions, especially when the bulb contains air at atmospheric pressure; and the coating is not always complete. It may be that these deficiencies are asso- ciated with some kind of "soiling" of the wall, but the available data do not justify any positive statement re- garding that. In one case it has been recorded that after the frost had been melted it reformed, beginning at or near the frozen water (− 9°C) and growing upward.

(2) *Ice.*—When a flask containing water at 0°C was plunged into a cold bath (− 12°C), ice formed first on the portion of the wall that was above the waterline, and a little later, on the surface of the water.

A stoppered and capped quartz bulb containing water freezing at − 11.6°C was observed at − 10.0°C to have patches of ice on its wall above the level of the (unfrozen) water in its bulb. When the bulb was tipped, the water froze as soon as it touched the ice. The distance between the lower edge of the wall-ice and the undisturbed meniscus of the water was not recorded.

On another occasion a star of ice 2 mm in diameter, quite like a snow crystal, was observed on the wall of a bulb at 10:13 A.M., temperature − 10.3°C. Its center was about 12 mm above the meniscus of the unfrozen water in the bulb. At 10:15 its diameter was 4 mm; at 10:18, temperature − 9.1°C, it was 8 mm; and at 10:31 (− 7.1°C) it touched the meniscus, and the entire volume of water froze at once. There were three small discrete drops of water on the wall and in the path of the growing star of ice. As the ice ap- proached it, each drop progressively decreased in size, and vanished before being reached by the ice. It seems plain that the growth of the star was by a reverse sublimation, by the capture of vapor molecules by the ice.

(3) *Propagation along the Wall.*—Early in the work, two bulbs, joined by their necks to form an inverted V, were charged with water, exhausted, and sealed. One was then charged from the other by low- temperature distillation without ebullition. When pro- gressively chilled, both specimens froze at the same time at − 12°C. But after one bulb had been sealed off from the other, the contents of one (the residue) froze at − 12°C, as before, and that of the other (the distillate) at − 16°C. While they were connected, the freezing of the residue at − 12°C presumably in- itiated that of the distillate by the propagation of ice through the connecting film of water on the wall.

At that time it was thought that such propagation necessarily occurred. Several similar observations seemed to justify that belief. But others showed that such propagation need not occur; and frequently does not. For example, such propagation did not occur in the case of the stoppered and capped quartz bulb nor of the star of ice, mentioned in the preceding subsection.

And, although but scant attention was given to the phenomenon, in certain cases it has been recorded that the drop of water, which frequently hangs in the neck of a bulb, remained unfrozen (see fig. 28), al- though the bulk of the liquid froze suddenly in the usual manner.

Still another illustration of the same thing is af- forded by the following incidental observations on diphenyl ether—$(C_6H_5)_2O$. To a colleague, Mr. E. F. Mueller, I am indebted for a sealed Pyrex tube about 30 cm long by 1.5 cm in diameter, filled within 4 cm of its top with diphenyl ether. A little of the ether was

FIG. 28. Unfrozen drop in neck of bulb. The photograph was taken while the bulb was still in the cold alcohol.

frozen in one end of the tube, and by suitable cooling, small clear crystals with sharp advancing edges were grown from that along one side of the wall. The tube was then placed on a suitable stand and inclined until the meniscus was within 2 or 3 mm of the tip of the most advanced of those crystals, which were on the lower side of the tube. By means of a fixed microscope with an engraved ocular micrometer, the distances from the tip of the crystal forward to the meniscus and back to a well-defined mark on the crystal mass were measured from time to time. Measurements in the two directions checked satisfactorily. Three sets of measurements, extending over 100 to 150 hours each, were made. The mean temperatures for the sets were 22.6, 24.7, and 25.1°C (m.p. 26.9°C), and during any one set the room temperature varied one or 1.5°C. The crystal grew, linearly with the time, solely by reverse sublimation, and very slowly: 2.2 μ/hr at 22.6°C, 3.5 μ/hr at 24.7°C, and 4.5 μ/hr at 25.1°C. It is interesting to notice, though perhaps, in view of the uncertainties in the temperatures, of no significance, that the rate of growth increased with the mean temperature. In no case was freezing propagated along the wall.

The subject should be studied with greater care, but these observations indicate that freezing cannot be propagated along a wall film that is very thin, and that such thinness can result from mere drainage, even in the presence of saturated vapor.

5. HETEROGENEITY OF WATER

If the thermal initiation of freezing in the water used in this study is due to the presence of motes, as herein suggested, then such waters are by their very nature heterogeneous, and it becomes of interest to determine the extent of that heterogeneity; that is, to determine how large a specimen must be in order to

TABLE 9

HETEROGENEITY OF WATER: ILLUSTRATIVE

Unless bracketed with others, each tabulated value of the spontaneous-freezing-point refers to a different specimen of water. All specimens in a given group were taken from the midlayers of the same large volume of water.

Spontaneous-freezing-points in °C. 100 drops = 3 ml.

Water	Distilled-1		Cold-faucet		Asbestos	
Bulb	G	H	A	C	C	D
Volume	120 drops		60 drops		120 drops	
Remarks	Settled	Shaken			Unfiltered, milky	
	−13.3	−12.0	−10.8	−9.1	− 6.8	− 7.0
	−12.9	−12.7	− 7.9	−7.1		
	−13.2	−12.0	− 7.0	−6.9	Filtered	
	−13.3	−12.0	− 7.0	−9.2	−11.7	−12.0
	−13.3	−12.0		−8.7	−11.4	−12.0
				−9.0	−12.6	−12.1
Water	Distilled-2		Cold-faucet		Asbestos	
Bulb	C	D	C	D	C	D
Volume	60 drops		10 drops		120 drops	
Remarks					Clear, unfiltered	
	−14.0	−14.7	−12.6	− 9.6	− 8.5	−13.0
	−14.0	−13.6	−12.7	− 9.8	−11.1	− 8.6
	−14.9	−14.9	−13.1	− 8.6	−10.3	−12.0
	−14.4	−14.3	− 8.0	−12.0		
	−14.3	−14.3				
Volume	10 drops		20 drops		120 drops	
Remarks					Bottom and sediment	
	−12.0	−15.1	−7.2	−7.0	−6.0	−6.0
	−15.1	−14.2			−6.7	−6.0
	−13.7	−15.0			−6.0	−6.0
	−14.0	−15.3				

be a fair sample of the volume from which it was taken, and how great may be the variations among several smaller samples, all of the same size and from the same source.

Only a few observations bearing on the subject have been made, but they show that significant differences may exist between volumes as large as 3 or 4 ml, and that variations of several degrees in the spontaneous-freezing points exist for smaller samples. Illustrative data are given in table 9, where the observed values of t_{sf} are tabulated for specimens of various sizes and for several waters. Each tabulated value is for a different specimen unless it is bracketed with others, then all values embraced by the bracket refer to the same specimen, being successive determinations. Between chargings, the bulbs were well rinsed with water freezing near $-15°C$. The data for the water designated as "Distilled-1" show the reproducibility of successive determinations, and also show that shaking the water before drawing the specimen caused a rise of $1°C$ in t_{sf}. Each specimen was drawn from the midlayers of the bottle.

Those for "Distilled-2" show that variations between specimens as large as 60 drops (1.8 ml) are not very great, but between those as small as 10 drops there may be variations as great as $3°C$. The water from the cold-water faucet is still more heterogeneous. The values of t_{sf} for specimens of asbestos water (water shaken up with ignited asbestos fibers), unfiltered but clear, as large as 120 drops (3.6 ml), may differ by $4.5°C$; and the bottom layers of the same water, including sediment, freeze $6°C$ higher than does its filtrate.

Obviously, it is very unsafe to assume that the t_{sf} of each of two or more specimens drawn from the same volume of water will be the same, unless the specimens are at least as large as 3 or 4 ml.

Some interesting observations and remarks on the colloidal impurities that may exist in distilled water have been published by Pennycuick and Woolcock [65].

Observations indicating that carefully prepared liquids may be distinctly heterogeneous had previously been reported by others who have studied freezing. In some cases [36, 68, 84, 95] the significance of those observations had been recognized; in others [49, 58, 70, 74] it had not. And most unfortunately, previous observers, ignoring that evidence, have frequently used in their studies small samples, often much smaller than 1 ml [18, 24, 36, 58, 70, 74, 77, 78, 84, 85, 95].

Furthermore, the quantity studied by those observers has been, in general, the time required for a specimen to crystallize, as counted from the instant at which it was placed in a bath of known temperature. But if those substances behaved as water has been found to do in this study, then it is the temperature attained by the specimen, not the time required to reach it, that determines whether the specimen will freeze or not. If the several specimens of the substance were initially at the same temperature and were cooled at the same rate and so slowly as to obviate complications arising from the existence of a finite interval of time between the initiation of freezing and the instant at which the observer becomes aware of it, then the specimen having the highest t_{sf} would crystallize first; and that having the lowest, last, unless the t_{sf} of such specimen were below the temperature of the bath, in which case the specimen would not freeze at all.

6. REMOVAL OF MOTES

Many methods for removing motes from water have been proposed and studied, mainly by those concerned with the preparation of optically empty water [37, 48, 53, 54, 65, 66, 93]. It should, however, be remembered that the optical requirements are much less severe than those for the present type of work. For the former, all that is required is that the total number of motes remaining shall be very small; for the latter, there must not be a single mote that is effective above the temperature to which it is desired to supercool the melt.

In the course of this work, several tricks have been tried after a rough-and-ready manner, the most satisfactory being the well-known procedure of low-temperature distillation in a vacuum, without ebullition. It is necessary to rinse the collector vessel well with the distillate, and to pour this rinse-water back into the reservoir without opening the system, so as to wash from the walls of the collector such motes as may have been adhering to it. That should be repeated several times before the collector is finally charged with distillate for freezing. And in order to avoid casual contamination, the collector should be hermetically sealed before being separated from the reservoir.

Obviously, this procedure should not be expected to remove all the motes. Some of every size will escape from the liquid, and some of them will be carried over by the vapor. The distillate may be expected to contain the same kinds of motes as the residue, but fewer of them; and the larger motes may be expected to pass over less readily than the smaller ones.

Furthermore, the lower the velocity of the stream of vapor, the less likely is a large mote to be carried over with the vapor. And the presence of an inert gas through which the vapor has to pass by diffusion will greatly impede the passage of motes. The first is secured by using a small difference in temperature between reservoir and condenser; the second, by distilling under reduced air pressure, instead of in a vacuum (actually in water vapor saturated at the existing temperature). A single trial with reduced air pressure was encouraging, but distillation was very slow.

Since the distillate is nearly pure water, whereas the residue may be a nearly saturated solution of the motes, it may be advantageous to heat the distillate for some hours, so as to hasten the solution of such motes as it may contain, and then to redistill from that. Such redistillation has been tried (cf. fig. 21), but not under satisfactory conditions.

In an earlier section ("Distillates and residues") may be found what seem to be illustrations of the carrying over of motes and of their later solution; see especially the graphs for CIV and C7 (fig. 22).

Still another effect may be observed if the distillation results in a relatively great change in the volume of water in the reservoir, as when the reservoir is of the same size as the collector bulb to be charged. If the initial water is a nearly saturated solution of the motes, distilling away half of it may cause deposition of the solute upon the motes present, thus increasing their size and changing the t_{sf} of the residue, raising it above that of the initial water if the motes are minor ones. If the wall film is thick enough to transmit freezing from one bulb to the other, and if the combination is slowly cooled until freezing occurs, the two specimens will seem to freeze at the same time, at the t_{sf} of the residue.

That seems to be the explanation of the following observations: The necks of two bulbs were connected by fusion so as to form an inverted Y, the system was charged with nearly enough distilled water to fill one bulb, the air was pumped out, and the stem of the Y was hermetically sealed. The water was then equally divided between the two bulbs, the system was placed in a chilling bath, and the temperature was lowered slowly until the water froze. Both froze at $-14.1°C$. Then all the water was poured into one bulb, designated as m, and distilled into the other (b), washing back and repeating a number of times; finally b was half filled by distillation. Now both froze at $-13.2°C$. The distillation process was repeated a number of times, freezings being made at intervals; and presently with b half filled with distillate, both froze at $-11.8°C$. Then each bulb was sealed off from the stem of the Y and frozen. The residue, in bulb m, froze at $-12.0°C$; whereas the distillate in b froze at $-16.0°C$. Although the initial water froze at $-14.1°C$, the residue, having half its volume, presently froze at $-12.0°C$ and the distillate at $-16.0°C$. Three weeks later the freezing point of the residue had risen to $-11.0°C$, and that of the distillate had fallen to $-17.2°C$; presumably further deposition had occurred in the former case, and solution in the latter. The observed t_{sf} for m remained the same (around $-11°C$) for the next 5 years, whereas after 2 years that for b had risen to $-11.8°C$, and 3 years later it had fallen to $-13.9°C$. Four years later, the t_{sf} for m was still about $-11°C$; whereas that for b had again risen to $-11.7°C$.

A gradual solution of major motes probably caused those variations in t_{sf} for b, and may have vitiated the interpretation here given of the other changes. Only after the first draft of this portion of the report had been written did the author become fully aware of the importance of such motes. Nevertheless, he thinks it very likely that the interpretation here given is, in the main, correct. Many repetitions of the experiment will be needed to settle the question.

II. MECHANICAL INITIATION OF FREEZING

1. INTRODUCTION

While it is generally acknowledged that freezing may be initiated by various mechanical means, there appears to be no generally accepted, clear explanation of how it happens; and the observations heretofore reported have generally been made for the purpose of showing that such means may be efficacious, rather than for clarifying either the reason for it or the cause of the frequent failures. Indeed, the observers seem, in the main, to

have been so overwhelmed by the idea that the initiation of freezing necessarily involves an uncontrollable element of chance that they have remained unimpressed by the failures. That is unfortunate. It has prevented a careful consideration of the failures, and has led the observers to attribute to their observations a generality which is unjustified.

Most of the observations now to be reported were made solely for the purpose of seeing what actually happens; and in the main they were made before any theory of such freezing had been developed. They were dictated in large part by the observations and statements of earlier workers; and they began with those intimately related to the study of the supercooling of water. It will be seen that none of them conflict with the new theory of freezing given presently in this report, and that most of them might have been foreseen by means of it.

It is convenient to recognize three distinct types of mechanical disturbances that may initiate freezing, or that have been thought to do so: (1) Mechanical agitation, including pouring, splashing, and squirting; (2) Impact; and (3) Rubbing of one surface over another.

2. MECHANICAL AGITATION

It has been frequently stated that mechanical agitation of any kind is inimical to the supercooling of water, and the conclusion has been drawn that the agitation itself in some way initiates the freezing [30, 36, 58, 62, 70]. Were that true, it would be a matter of great importance in any experimental study of the supercooling and freezing of water. Consequently the subject was investigated at the very beginning of the present work—first by gently pouring the supercooled water over the inner wall of the bulb, and then by splashing it. And near the end of the work the possibility of squirting supercooled water was briefly investigated.

a. Pouring

It was found that, in general, the supercooled water can be poured over the entire inner surface of the bulb without the initiation of freezing. For example, the water in bulb CI at $-14°C$ was poured over the entire interior of the bulb without its freezing, although at that time its temperature of spontaneous freezing was only a little lower (about $-14.5°C$). But in a number of cases freezing did begin when the water was poured; and sometimes freezing would occur at an abnormally high temperature when, by tipping a bulb, the meniscus was caused to change its position very slightly, the motion being exceedingly gentle. Such exceptions must be accounted for.

That there are cases in which supercooled water very near its t_{sf} can be poured without its freezing is proof that the mere mechanical agitation incident to the pouring is not what initiates the freezing that occurs at significantly higher temperatures in the exceptional cases. In those cases the pouring can do no more than to call into play something that otherwise would have been inoperative. Indeed, when such freezing occurs as a consequence of a slight tipping of the bulb, the intensity of the agitation is essentially zero; the meniscus has merely moved smoothly over small areas of the wall of the bulb, in some places covering with water an area previously bounding the gas phase; in others doing the reverse. It is that motion over the wall to which one must look for the explanation of the freezing, but not to the motion itself; for in most cases

TABLE 10

EFFECT OF POURING

All temperatures are °C; t_{sf} = spontaneous-freezing-point; No (Yes) = water did not (did) freeze when poured at the indicated temperature; T = bulb tumbled the number of times indicated before the water froze; Chat. = the tumbling mechanism chattered as the bulb was turned; Quick = water froze within a few minutes after the bulb was placed in the chilling bath, probably before the water had reached the temperature of the bath; Erect, Hor., Inv. = bulb at rest and erect, horizontal, inverted, respectively. In the first section are data for 4 of the sealed bulbs; in the second are those for 6 of the stoppered and capped (SC) bulbs—those designated by Roman numerals are of vitreous quartz; those by Arabic are of Pyrex. These SC bulbs were cleaned and charged in the dusty room occupied after 1941. Bulbs SCI and SC1 were each used with three samples of water; the data for the several samples are separated by rules. The temperature under "No" is that which was read, and at which the bulb was tumbled, next before freezing occurred at the other temperature recorded on the same line.

As the bulb was tumbled the water poured from one end of the bulb to the other and back again, filling and emptying the neck. The pouring was never violent.

Hour	Temperature		Hour	Temperature		Hour	Temperature		Hour	Temperature	
	t_{sf}	No		t_{sf}	No		t_{sf}	No		t_{sf}	No
	C7			C18			C23			G3	
1:35	−13.4	—	4:22	−21.3	—	2:22	−5.8	—	4:00	−16.3	—
:52	−13.4	—	:36	−21.3	—	:26	—	−4.9	:12	—	−16.0
2:10	—	−13.0	:50	—	−19.8	:40	−5.9	—			
:15	—	−13.5									

Hour	Temperature			Remarks	Hour	Temperature			Remarks
	No	Yes	t_{sf}			No	Yes	t_{sf}	
	SCI					SC1			
12:15	−8.1	− 8.4	—	—	4:20	−10.6	−10.8	—	—
:35	−9.1	—	− 9.2	—	:40	−10.9	—	−10.9	—
2:15	−7.5	− 7.5	—	Several T.	2:55	−10.1	—	−10.6	—
3:20	−9.8	—	− 9.9	—	3:05	− 9.5	− 9.1	—	Quick
:35	−9.2	− 9.8	—	—	:25	−10.1	—	−10.4	—
:55	−9.8	—	− 9.8+	—					
4:07[a]	−8.8	− 9.0	—	2 T. Chat.	12:04	− 7.4	− 7.2	—	—
:29	−9.8	—	− 9.9	—	:59	− 9.3	—	− 9.5	—
:50	−9.8	− 9.9	—	—	2:46	− 7.8	− 7.9	—	—
					3:07	− 8.5	− 8.6	—	—
11:55	−8.1	− 8.3	—	—	:32	− 8.8	—	− 8.9	—
12:15	−6.8	− 7.0	—	—	:56	− 7.9	− 8.0	—	—
:42	−8.0	− 8.1	—	—	4:06	− 7.7	− 7.8	—	—
2:24	−6.8	− 7.0	—	—	:51	− 9.9	—	−10.1	—
3:43	−9.8	—	− 9.9	—	10:07[a]	− 8.4	− 8.5	—	—
4:41	−9.6	− 9.8	—	—	:15	− 7.5	—	− 7.8	Quick
9:45[a]	−7.0	− 7.2	—	—	:36	− 8.1	—	− 8.2	—
10:52	−9.7	—	− 9.9	—	11:23	− 7.7	− 7.9	—	—
11:05	−7.1	− 6.7	—	—	:38	− 7.9	—	− 8.1	—
:51	−9.8	−10.0	—	—					
					2:41	− 7.9	—	− 8.0	—
2:12	−5.9	− 6.2	—	—	3:26	− 9.1	—	− 9.3	—
:41	−7.9	—	− 8.0	—	4:04	− 9.5	—	− 9.8	—
:57	−7.9	—	− 8.0	—	3:00[a]	− 7.6	− 7.8	—	—
3:14	−7.9	—	− 7.9	—	:32	− 8.0	− 8.0	—	—
:48	−7.7	—	− 7.9	—	4:12	− 9.9	—	−10.0	—
					:30	− 9.8	−10.0	—	—
					:49	− 9.0	—	− 9.1	—

TABLE 10—*Continued*

Hour	Temperature			Remarks	Hour	Temperature			Remarks
	No	Yes	taf			No	Yes	taf	
	SCII					*SC2*			
10:35	−7.0	− 7.4	—	Jarred	10:48	− 7.5	—	− 7.8	—
11:35	−8.2	—	− 8.3	—	11:01	− 7.0	− 7.1	—	—
12:22	−8.1	− 8.2	—	—	:26	− 7.8	—	− 7.9	—
2:23	−7.4	—	− 7.7	Inverted	:49	− 7.0	− 7.1	—	—
:39	−7.5	− 7.7	—	—	12:03	− 7.0	− 7.0+	—	First T.
					2:05	− 7.6	—	− 7.8	—
					:39	− 7.5	—	− 7.3[b]	—
	SCIII					*SC3*			
11:25	−9.4	—	.− 9.8	—	11:42	−10.0	—	−10.2	—
:34	−9.2	—	− 9.7	—	:56	−10.0	—	−10.1	—
12:23	−9.1	− 9.2	—	—	12:11	− 9.0	− 9.1	—	—
:33	−8.9	− 9.0	—	—	:45	− 9.7	− 9.8	—	—
2:10	−7.4	—	− 7.9	—	3:59	− 9.8	− 9.9	—	—
:22	−8.2	− 8.3	—	—	4:12	− 9.6	—	− 9.8	—
:38	−7.3	− 7.3	—	—	:37	− 9.3	—	− 9.8	—
:53	−7.1	− 7.1	—	—	:56	—	—	− 9.3	Hor.
3:28	−9.0	− 9.0	—	—	10:45[a]	—	—	− 9.3	Hor.
:49	−9.1	—	− 9.2	—	11:08	—	—	− 9.4	Hor.
4:30	−9.1	− 9.4	.—	2 T.	:28	—	—	− 9.1	Inv.
:56	—	—	− 9.3	Hor.	12:08	—	—	− 9.5	Erect
10:45[a]	—	—	− 9.3	Hor.	:45	—	—	−10.2	Erect
11:08	—	—	− 9.4	Hor.	2:26	—	—	− 9.1	Erect
:28	—	—	− 9.2	Inv.	:54	—	—	−10.7	Erect
12:17	—	—	−10.1	Erect	3:44	—	—	−10.6	Hor.
:33	—	—	− 9.5	Erect	4:04	—	—	− 9.4	Hor.
2:26	—	—	− 9.1	Erect	:42	—	—	−10.4	Hor.
:44	—	—	− 9.1	Erect	5:16	−10.0	−10.1	—	—
3:31	—	—	− 9.3	Hor.	11:30[a]	—	—	−10.5	Erect
4:04	—	—	− 9.4	Hor.	:54	—	—	−10.4	Erect
:30	—	—	− 9.5	Hor.	12:15	—	—	−10.3	Erect
5:02	−8.8	− 8.9	—	—	:46	—	—	−10.2	Inv.
11:30[a]	—	—	−10.5	Erect	3:02	−10.2	−10.2+	—	4 T.
:42	—	—	− 9.8	Erect	:16	− 7.6	− 7.8	—	1 T.
12:07	—	—	− 9.4	Erect	4:12	−10.1	−10.2	—	—
:34	—	—	− 9.3	Inv.					
2:28	−8.3	—	− 8.8	—					
:47	−8.9	—	− 9.0	—					
3:26	−7.7	− 7.5	—	1 T.					
:53	−8.8	—	− 8.9	—					

[a] Another day. The interval may be merely overnight, or over the week-end.
[b] The bulb had been as cold as −7.7°C with the water still unfrozen.

such motion is not accompanied by any freezing. The obvious explanation is that there is something on the wall in the gas phase that initiates freezing when the supercooled water is brought into contact with it. That something may be either a minute crystal of ice (it has already been shown that the wall film in the gas phase may be so thin that the process of freezing cannot be propagated through it) or a mote. (Whether the mote merely adheres to the wall or is a part of it —a point or scratch, etc.—is a matter of indifference.) The same explanation is available in other exceptional cases, and readily accounts for the great variability in the observations. In some trials, motes or crystals are present on the walls; in others they are not, the motes having been washed away and the conditions necessary for the formation of crystals having been removed.

Typical data are given in table 10. In the first section are given data for 4 of the sealed bulbs. At the temperature recorded under "No" the specimen did not freeze when the bulb was lifted from the alcohol bath, quickly grasped by its neck with a precooled wooden clothes pin, and at once tipped so as to pour the water over the portions of the wall that had been in contact with the gas phase. Even when the temperature was very near the t_{sf} of the specimen the water did not freeze. In the other sections of the table are data for 6 stoppered and capped bulbs that had been cleaned and charged in the dusty room occupied after 1941. By a suitable device those bulbs could be tumbled end over end without removing them from the bath.

Observations were made in this manner: At one-minute intervals the temperature of the bath was read and then a little solid CO_2 was added. Immediately after the addition of the CO_2 the bulbs were tumbled once or twice; and about a quarter of a minute before the next temperature was read, they were again tumbled so as to pour the water into the neck of the bulb and out of it again. In this table are given under "No" the lowest recorded temperature preceding the freezing, and the temperature of freezing is recorded under "Yes" if freezing occurred while the bulb was being tumbled, and under t_{sf} if it occurred while the bulb was at rest. In some cases in which there was no freezing when the bulb was tumbled, there was freezing when the tumbling was repeated a number of times in quick succession. Whether that indicates that there is an appreciable delay between the tumbling and the appearance of freezing, or that the effect of tumbling is cumulative, or that the effect arises from the slight progressive fall in the temperature of the water, can be determined only by additional investigation. But there have been observed a number of things hard to record in a significant manner that suggest that there is, at least in some cases, an appreciable, though usually very short, delay between the inception of the process of freezing and the appearance of ice. That there should be such a delay should cause no surprise.

As may be inferred from the tabulated values many determinations must be made if the observer would assess fairly the effect of changes taking place in the specimen—of changes in the sizes of the motes.

b. Splashing

Early in the work it was found that, when repeated pourings of the supercooled water over the wall of the bulb caused no freezing, a violent splashing of the water immediately thereafter might, and usually did, cause freezing, although the water was then somewhat less supercooled than when the pouring was begun. That effective splashing was done in this manner: With the neck of the bulb grasped firmly between the thumb and index finger, bulb vertical and neck up, that hand was struck violently downward against the other; the splashing had to be violent, so as to shatter the mass of water, and the temperature had to be low. The higher the temperature above the t_{sf} of the specimen, the more violent the splashing had to be; and, as for tumbling, there were indications that the effect of splashing might be cumulative—several in quick succession causing freezing when one would not. Although observations to determine the relation between

the violence of the splash and the highest temperature at which it is effective for a given specimen might give valuable information, none have been made in the course of this study. The data given in table 11 are typical. They seem to indicate that, if the temperature is more than a degree or two above the t_{sf} of the specimen, it is not easy to initiate freezing by splashing.

TABLE 11

INITIATION OF FREEZING BY SPLASHING

Blb = bulb; Hr. = time of day at which the temperatures were read; t_{sf} = spontaneous-freezing-point; No (yes) indicates that splashing did not (did) induce freezing at the indicated temperature; Δt = excess of the other temperatures above that of t_{sf}.

Blb	Hr	t_{sf}	No	Yes	No	Yes
		Temperature			Δt	
C49	9:41	− 9.0	—	—	—	—
	:46	—	− 7.2[a]	—	1.8	—
	:52	—	—	− 8.6	—	0.4
SCI	3:47	− 9.9	—	—	—	—
	4:08	—	—	− 9.0	—	0.9
	:30	− 9.9	—	—	—	—
C8	2:51	−10.8	—	—	—	—
	3:05	−10.8	—	—	—	—
	:14	—	− 9.4	—	1.4	—
	:18	—	—	−10.1	—	0.7
	:23	—	—	−10.0	—	0.8
SCII	11:40	—	—	−10.2	—	1.4
	12:14	−11.6	—	—	—	—
	:26	−11.7	—	—	—	—
C7	1:36	−13.4	—	—	—	—
	:52	−13.4	—	—	—	—
	:59	—	—	−11.9	—	1.5
	2:15	−13.5	—	—	—	—
CI	3:02	−15.0	—	—	—	—
	:16	−15.0	—	—	—	—
	:22	—	—	−13.7	—	1.3
	:31	—	—	−14.0	—	1.0
	:38	—	—	−14.0	—	1.0
C17	4:27	−15.1	—	—	—	—
	:40	−15.2	—	—	—	—
	:47	—	—	−14.1	—	1.0
	:59	—	—	−14.0	—	1.2
G3	4:02	−16.3	—	—	—	—
	:12	—	—	−16.0	—	0.3
C18	4:22	−21.3	—	—	—	—
	:36	−21.3	—	—	—	—
	:50	—	—	−19.8[b]	—	1.5
	5:04	—	−19.9	−19.9[c]	1.4	1.4
	:14	—	−20.0	−20.0[c]	1.3	1.3

[a] Shook violently.
[b] After third splash.
[c] Many gentle splashes did not induce freezing; a violent one did.

Early in the work I obtained the impression that cavitation was probably in some way responsible for the initiation of freezing by splashing. No observations made in the course of this work suffice to establish that supposition, but it seems to accord with the theory gradually developed during the course of this work. The rupture involved in cavitation may well produce an effect that is similar to that produced by molecular impacts in the thermal initiation of freezing. And it is possible that minute motes may be torn from the glass when the liquid is splashed. Füchtbauer [36] has suggested such a tearing, but he attributed the freezing to the resulting formation of a fresh glass surface, of one not covered with an adsorbed layer of the melt.

These explanations of the effect of splashing inevitably raise another question. Will the steep transverse gradient in the velocity that attends a very rapid flow of water over a solid produce a similar effect, and so lead to freezing? The experiments on the squirting of supercooled water, to be reported in the next section, were intended to throw light on that question. But there should first be recorded certain observations on the effect of splashing aqueous solutions of ethyl alcohol (about 14 per cent of alcohol by weight). A solution freezing spontaneously at $-25°C$ was splashed by hand at $-13.4°C$. It froze only after several splashes, and the first visible sign of freezing was the appearance of discrete pieces, or balls, of ice disseminated throughout the liquid. Another solution, freezing spontaneously at $-21°C$, was splashed at $-20°C$ in a seesaw machine that permitted marked splashing with a very small motion of the bulb. After the third splash, a single small rosette appeared and grew. In another trial, seesaw splashing, though repeated many times, caused no freezing, but did give rise to many air bubbles distributed throughout the liquid. Hand splashing immediately thereafter gave rise to numerous rosettes in or near the free surface of the liquid. In another trial, seesaw splashings produced an emulsion-like mixture of air and liquid, hand splashing increased the amount of that emulsion, which seemed to be markedly viscous; after several splashings by hand, the liquid froze.

None of these phenomena have been observed with water, and such tests have not been made with any other solution.

c. Squirting

Can a high-speed jet of supercooled water be squirted from a glass nozzle? Or will the disturbances incidental to the squirting cause freezing at the very beginning of the process?

As there seemed to be no recorded experiments bearing on the subject, and a rather general feeling that freezing would surely occur, it was decided to make a few tests, solely for the purpose of seeing whether freezing necessarily occurs under those circumstances.

Since it is necessary for the air through which the jet is squirted to be cooled below 0°C, and since that was done by means of solid carbon dioxide, it is probable that the air usually contained minute crystals of ice. If any one of those crystals were to come into contact with the wet nozzle or with the solid jet of water proceeding from it, freezing would occur, blocking the nozzle, and would rapidly extend to the reservoir of water. Consequently, definite conclusions can be drawn only if it be found possible to squirt the water. Many failures are to be expected.

Nevertheless, by air pressure, water at $-3°C$ has been squirted through a glass nozzle in a fairly flat trajectory through air at $-4°C$ to impinge on a test tube 25 cm away containing alcohol at $-4°C$. Although there were three distinct, but closely following, periods of squirting, the water did not in any case freeze in the nozzle, but did freeze when it hit the test tube, building up on that a projecting mass of ice similar to those often formed on the leading faces of projections on airplanes. In another trial, water at $-5°C$ was successfully squirted through air at $-2°C$. Several such successful tests were interspersed with many failures.

Whence it is obvious that the disturbances involved in such squirting of supercooled water—high transverse velocity gradient, the production of a new air-water interface just beyond the nozzle, and the breaking of the jet into drops—need not of themselves initiate freezing even when the streaming velocity is high.

Also the fact that a mass of ice built up on the cold test tube, although freezing did not progress backwards from that along the jet to the nozzle, suggests that the jet had broken into discrete but unfrozen drops before it reached the test tube. Additional experiments, however, to prove that the failure of such backward progression did not arise on account of the high forward velocity of the jet, are needed before that conclusion can be unreservedly accepted.

3. IMPACT

Previous observers [15, 94, 97] have reported that freezing may be induced in supercooled water by the mechanical disturbance or shock produced by the impact of one solid on another. In general the solids have been immersed in the water. In those early observations, freezing certainly accompanied the impact. But that it was induced by the mechanical disturbance or shock itself is not so sure. It may have resulted from an incipient cavitation caused either by the rebound of the hammer or by the shock wave caused by the impact. And in every case there was motion of the meniscus, and that may have brought the water into contact with an effective mote adhering to the wall above the undisturbed water-line; then that mote would have initiated the freezing irrespective of any other effect of the disturbance. Also it is exceedingly difficult to avoid all rubbing of the impinging surfaces on one another, and when that occurs within the water, the rubbing alone may, as will presently be shown, induce freezing. As none of these factors seemed to have been given sufficient attention in the earlier work, it was decided to study the subject anew, bearing them in mind.

In order to avoid any possible effect of rubbing, the first experiments consisted in subjecting some of the sealed bulbs of water to external blows, jars, and mechanical shock of various kinds. The mere mechanical effect of such treatment upon the water just inside the bulb will differ but little from that of corresponding impacts upon the inner wall of the bulb. There will, of course, be a motion of the meniscus, but should it be found that in a reasonable number of cases the mechanical shock fails to induce freezing, then one would be justified in concluding that the shock itself is not effective, that one must look elsewhere for the explanation of the freezings that do occur. That proved to be the case.

It was found that a much supercooled bulb of water could be struck violently with a similarly cooled rod while immersed in the chilling bath without causing the water to freeze. The same result was obtained with water cooled in an open test-tube. And a bulb containing water that froze spontaneously at $-16.3°C$ could be rapped sharply when at $-16.0°C$ without the water's freezing. At another time, sealed bulbs of supercooled water were attached to a steel rod (13 mm in diameter by 20 cm. long) which was allowed to fall vertically through guides to strike a fixed steel block. The bulb was held vertically by its neck, clamped in a precooled wooden clothes-pin. A bulb containing water freezing spontaneously at $-16°C$ and cooled to -9 or $-10°C$ was thus jarred by dropping the rod through various distances up to 5 cm, but seldom did the water freeze, although the splashing was not slight. Also, the water in a supercooled bulb, clamped to the rod as for the dropping experiment, did not freeze when the rod, held in the hand, was struck sharp longitudinal blows with a hammer. That was repeated a number of times, always with the same result.

All these experiments show that the mere mechanical shock attending impact, up to the intensities here employed, does not suffice to induce freezing in supercooled water; at least it does not suffice at the temperatures and with the waters here employed.

Other experiments were made in which the impinging surfaces were immersed in the supercooled water. For those experiments the water was contained in a large test tube closed by a paraffined cork through which passed a central axial glass bushing which fitted snugly the stem of a thermometer passing through it. Water freezing spontaneously at $-8.5°C$ was cooled to within the range -4.5 to $-6.3°C$, and then, without removing the tube from the chilling bath, the thermometer was tapped against the bottom of the tube, beginning with a gentle tap and gradually increasing

the intensity as much as one dared, considering the safety of the thermometer. Many tests of that kind were made. Only very exceptionally did freezing occur. But if the thermometer were rubbed, ever so gently, against the tube, the water froze at once. Sliding of one surface over the other always caused freezing; tapping, with care to avoid sliding, seldom did, the exceptions probably arising from a slight unintentional sliding.

Mere stationary contact between the surfaces had no effect upon the temperature to which the water could be supercooled; and with the avoidance of sliding, a mere breaking of that contact was never accompanied by a freezing of the supercooled water.

Whence it is concluded that neither the mere mechanical disturbance nor the shock incident to the mutual impact of two solids nor the stationary contact of two solids immersed in the water is of itself efficacious in initiating the freezing of supercooled water under the conditions employed in these tests.

But even a very gentle sliding of one glass-water interface over another may be very efficacious. To the effect of that kind of disturbance, we now turn.

4. RUBBING

In any study of the initiation of freezing of supercooled water by the rubbing of one water-solid interface against another, three types of rubbing are to be distinguished: (1) the linear dragging of one surface over another; (2) the relative twirling of two surfaces of revolution about their common axis, the surfaces being so adjusted that the common axis of revolution passes through their point of contact; and (3) the twirling about its central normal of a circular plane in contact with a stationary surface, either a second plane or a surface of revolution, tangential to the twirling plane at its center.

The last has not been investigated during the study now being reported.

The second was investigated in this way: with the test tube and thermometer set up as described a few paragraphs earlier, the thermometer was twirled about its axis while its bulb was held firmly pressed against the center of the bottom of the test tube. Very seldom indeed did such twirling initiate freezing, the exceptions being probably caused by a slight unintentional sliding of the bulb along the bottom of the tube. I have no satisfactory explanation to offer for the inefficacy of such twirling under conditions in which the linear dragging of the surfaces over one another was very efficacious.

The first case, that in which one surface is dragged linearly over another, has been studied in some detail, with interesting results. The rest of this section will be concerned with that type of rubbing. Here also, many of the experiments were made before any theory had been developed to cover them; they were dictated primarily by curiosity, guided in part by statements that had been made by others. The experiments will be presented in their approximate chronological order so that the reader may the more readily see why certain of them were undertaken.

It was found that a very gentle wiping of a glass-water interface with a camel's-hair brush may initiate freezing as readily as a rubbing of it with another glass-water interface, the temperature of the water being a degree or two above its t_{sf}.

Late in 1938 the following observation was made. In a large Pyrex test tube containing water that froze spontaneously at $-8°C$ were suspended a thermometer and two glass rods coated with paraffin (probably Parowax). The thermometer and rods dipped deeply into the water but did not touch either one another or the tube. The whole was placed in a bath at $-7°C$, and when the thermometer in the water read about $-5°C$ the rods were rubbed one against the other while still immersed. That rubbing, whether the pressure were great or slight, did not initiate freezing. But a rubbing of either against the bulb of the thermometer did initiate freezing at once, even when the pressure was slight. And when the rod was removed from the thermometer as quickly as possible after the rubbing, freezing proceeded from it as well as from the thermometer, showing that the ice adhered to the paraffin as well as to the glass. This experiment was repeated a number of times, always with the same result.

The subject was not investigated further until the summer of 1942. It was the first experimental work undertaken after moving into the dusty room, and was begun prior to a full appreciation of the dustiness of the air.

At that time the new theory, presently to be proposed, had begun to take shape. It seemed to give a reasonable explanation of the mechanism by which rubbing initiates freezing, and to indicate that it would probably be impossible to initiate freezing by that means unless the melt were significantly supercooled. But it did not seem likely that the governing temperature would be at all well defined. The idea that there might be a temperature, fixed by the existing conditions within a comparatively narrow range, above which the initiation of freezing by ordinary rubbing would be impossible, had not presented itself.

Consequently, the early work with paraffin coated rods suggested that freezing could not be initiated by the sliding of paraffin over paraffin; and the work of 1942 began with a search for other substances resembling paraffin in that respect.

Twenty-two different substances, mainly waxes, greases, and oils, that happened to be readily available, were investigated briefly and with no attempt at real precision. It was found that in every case freezing could be initiated by rubbing a substance on itself if the temperature was as low as $-6°C$, and in no case

was freezing so initiated if the temperature was above $-2°C$.

Contrary to what was expected, these results indicated that there is no qualitative difference between the behaviors of paraffin and glass, and that the temperature does play a significant role.

Consequently, the search for substances that do not initiate freezing when they are rubbed together in supercooled water was discontinued, and plans were made for studying each of a few substances for the purpose of discovering such relations as may exist between the t_{sf} of the water and the highest temperature at which freezing can be initiated by rubbing the substance on itself and on glass, the reproducibility of that highest temperature (hereafter called the rubbing temperature), and factors that may modify it. In general, the unqualified term "rubbing temperature" (t_r) will refer to the rubbing of the substance on itself; for other cases, qualifying words will be used, but in every case one of the rubbing surfaces will be understood to be that of the substance under study. If the other surface is glass, t_{rg} will be used to indicate the highest temperature at which freezing is so initiated. Perhaps on account of the dustiness of the air in the room and of the resulting soiling of the relatively large area of the glass-gas surface of the interior of the vessel holding the water, the data are not satisfactorily reproducible. They do, however, suffice for certain purposes.

The following method and procedure were used. The water, about 35 ml, was contained in a 3 by 20 cm Pyrex test tube closed by a paraffined cork through which passed with convenient friction a thermometer and two glass rods, each rod carrying at its lower end a specimen of the substance under test. The shortest distance from the wall of the tube to the thermometer and to each of the rods was about 7 mm. One rod had a joint of spirally wound copper wire a short distance under the cork, and the bottom end of that rod was bent and suitably curved in a plane perpendicular to the length of the rod. The other rod was continuous and straight. By a suitable axial rotation of the bent rod, its bent portion (called the slider) could be brought into contact either with the other rod or with the bulb of the thermometer, and could be slid along it by sliding the rod through the cork. The slider could also be brought into contact with the surface of the tube, and be slid along it. The spiral wire joint was for the purpose of reducing the risk of breakage. In studying the effect of sliding glass over glass, clean rods were used; in studying the sliding of paraffin, the lower portion of the straight rod and the slider were each coated with the paraffin under study. Several kinds of paraffin were used.

The test tube with its contents was suspended in an alcohol bath, the bottom of the tube being immersed to a depth of about 14 cm; and the bath was cooled by additions of small amounts of solid CO_2 at 2-min intervals. The order of procedure was this: The temperatures of the bath and water having been read and recorded, the slider was touched to the vertical specimen (or to the thermometer) and at once withdrawn, then touched again and slid along it, then if there were no freezing it was withdrawn and small bits of CO_2 were added to the bath; two minutes after the preceding temperatures had been read, this process was repeated; and so on until freezing occurred. Frequently, both the thermometer and the vertical specimen were rubbed, one immediately after the other; and in some cases the wall of the test tube was also rubbed. For reasons that will appear, no data for the rubbing of the wall are given in this report. From time to time determinations were made of the temperature at which the water in the tube, as set up, froze spontaneously.

With such a large volume of water (35 ml) the time required for a determination is long unless the bath is kept significantly colder than the thermometer dipping

in the water. The average difference between those temperatures was between 1° and 2°C—in some cases as small as 0.5°C at the time of freezing, in others it was as great as 3°C. Such differences inevitably introduce uncertainty in the determination of t_{sf}, but, in view of the symmetry of the set-up, it seems likely that they will produce very little effect on either of the rubbing temperatures—specimen or thermometer. Determinations of t_{rg} for the wall were made in hope that a mutual study of it, of that for the thermometer, and of the temperature of the bath, might lead to an approximate determination of the amount by which the true t_{sf} lay, on account of the lower temperature of the bath, below the corresponding temperature of the thermometer in the water. On account of erratic variations, perhaps attributable in part to the dustiness of the air, and surely in part to inherent variations in the efficacy of the rubbing, that hope was not realized.

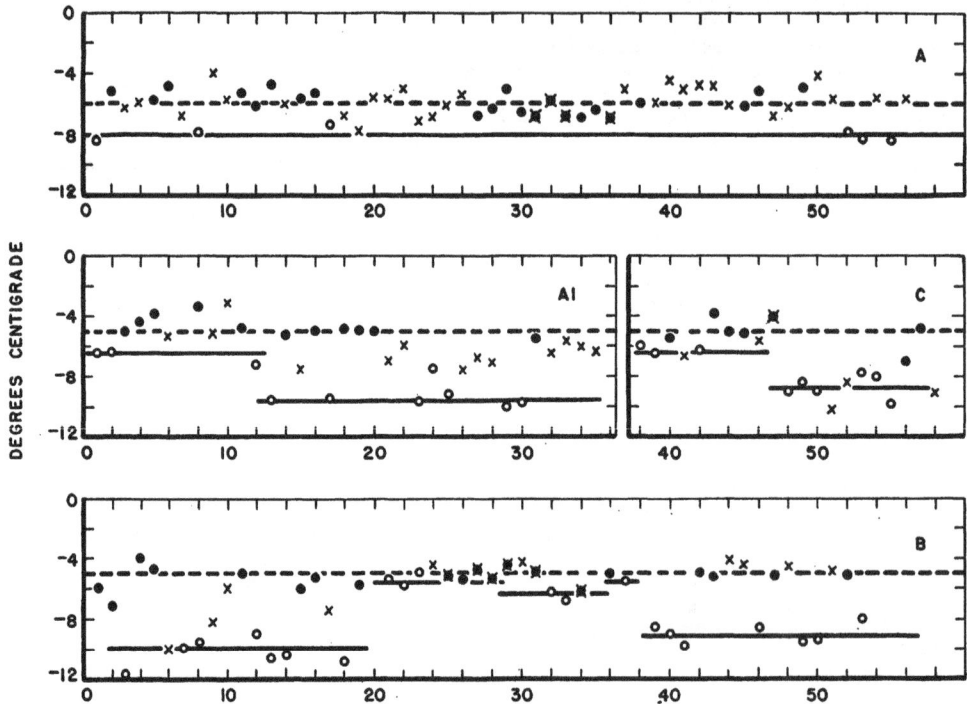

FIG. 29. Initiation of freezing by rubbing. Parowax fused to glass, except as noted. Ordinates, temperature at which freezing occurred; spontaneously, rings; on rubbing vertical specimen, crosses; on rubbing bulb of thermometer, dots; a dot in an open cross means that freezing occurred shortly after the specimen and thermometer had each been rubbed. Abscissas serve merely to separate the observations and to facilitate reference. Section A. Distilled water. Coated slider was 3-mm glass rod from 0 to 14, glass cat's whisker 15 to 56; from 40 to 56 vertical specimen was an uncoated glass rod, crosses as well as dots refer to rubbing of Parowax on glass. Section A1. Distilled water. From 1 to 6 slider uncoated, hence dots are glass on glass, crosses are glass on Parowax; after 12 water preheated in Pyrex for 8 hours in boiling water. Section B. From 0 to 20 same water as A; 21 to 28 and 36 to 38 D.C. faucet water; 29 to 34 other distilled water; 39 to 53 a third lot of distilled water, preheated 3¾ hours. Section C. Water as for A. After 47 the slider was recoated.

Consequently, the value recorded for t_{sf} in these determinations is the corresponding reading of the thermometer dipping in the water.

In no case did a mere tapping of the slider against either the vertical specimen or the thermometer initiate freezing.

Determination of t_r and of t_{rg} is frequently complicated by a delay between the rubbing and the appearance of the freezing to which it gives rise. That delay has not been timed, but in some cases it seemed to be several seconds. Furthermore, although several rubbings made in quick succession may fail to induce freezing, a single rub made a few seconds later may induce it, perhaps after a brief delay. That a difference in the energy expended is responsible for such effects seems incompatible with the data shown in A of figure 29. For abscissas 0 to 14, the rubber was a 3-mm glass rod coated lightly with Parowax; for the others it was a very slender glass cat's-whisker similarly coated. For the first, the pressure was not much under what the rod could stand; for the second it was usually just enough to show by the bending of the whisker that contact had really been made. The two sets of data substantially agree.

Of the four typical sets of data for Parowax, shown in figure 29 and obtained in the manner already indicated, it will be noticed that in A, in $A1$ 0 to 12, in C before 47, and in B after 20, there is essentially no distinction between the temperatures corresponding to the crosses (Parowax rubbing on Parowax) and those corresponding to the dots (Parowax rubbing on glass); whereas in the other portions the crosses are distinctly lower, and tend to become confused with the circles (t_{sf}). An actual confusion of those two would indicate that freezing could not be initiated in those waters by rubbing Parowax on Parowax—a conclusion that was initially drawn, on insufficient evidence, from the observation of 1938. The uncertainty as to the proper interpretation of these data cannot be resolved until many more such observations shall have been made under much better conditions, in a room with air that is much cleaner than that in which these observations were made. For where there seems to be no distinction between the crosses and the dots it is entirely possible that they do not represent the effect of rubbing Parowax on Parowax and on glass, respectively, but of motes on motes or of motes on glass, the surface of the Parowax, and possibly of the glass also, being thoroughly fouled by airborne motes, the fouling of the Parowax in the other cases being much less serious. Such fouling might also account for an obvious tendency, not only here but in many other cases, for the temperature corresponding to the crosses to be the same as that corresponding to the dots.

Another striking feature of figure 29, confirmed in many other cases, is the apparent independence of t_r (t_{rg}) and t_{sf}. It looks much as if the former might be determined solely by the rubbing surfaces, whereas the latter, if it does not coincide with the former, is determined solely by the specimen—by the motes in the water and on the walls of its container.

A short series of observations of the same kind as those shown in figure 29, but made with a pure paraffin ($C_{32}H_{66}$), very kindly supplied by my colleague Dr. F. D. Rossini, yielded results of the same general kind. Since that hard paraffin soon broke away from the glass when subjected to such variations in temperature, a long series could not be obtained.

5. EFFECT OF SOLUTES ON MECHANICAL INITIATION OF FREEZING

Those who have preferred to regard freezing as a volume effect, as consisting in its essence of a co-aggregation of molecular complexes (such as polymers) in water, point out that a very slight change in the solute content of the water—in particular, a very slight change in its hydrogen-ion content—may make a profound change in the amount of such preexistent complexes, and in that way may profoundly change all the phenomena that characterize freezing. Their reasoning seems to be sound. Consequently, it becomes necessary to determine experimentally how the phenomena now under study are affected by the addition of solutes to the water.

Data given and discussed in an earlier section indicate that the presence of dissolved substances in the water produces but little if any effect upon its t_{sf}; the effect not exceeding the corresponding depression of the normal melting point. Is the same true for the mechanical initiation of freezing?

No previous experimental study of this question seems to have been made; and in this study it has been done only for the case of rubbing. For that, many experiments were made. In some, paraffin coated rods were rubbed either on similarly coated rods or on glass; in others, uncoated rods were used, and the rubbing was of glass on glass. As in the work already discussed, the data for the coated rods were the more erratic. With that exception, there was no qualitative difference between the two cases. The procedure was to determine, in the manner already described, the temperatures at which the specimen of water to be studied froze (1) spontaneously and (2) when subjected to a sliding of one solid over the other. Then a very small amount of an aqueous solution was added, and those temperatures were again determined; then more solution was added; and so on. Solutions of HCl, NaOH, NaCl, NaNO$_3$, cane sugar, several combinations of HCl and NaOH added successively and in both orders, and the very active chromic cleaning solution (potassium bichromate and concentrated sulphuric acid), were used. A few sets of typical data for uncoated rods are shown in figures 30 and 31.

Of the solutes studied, only the very active chromic solution produced any significant change in the t_{sf} of

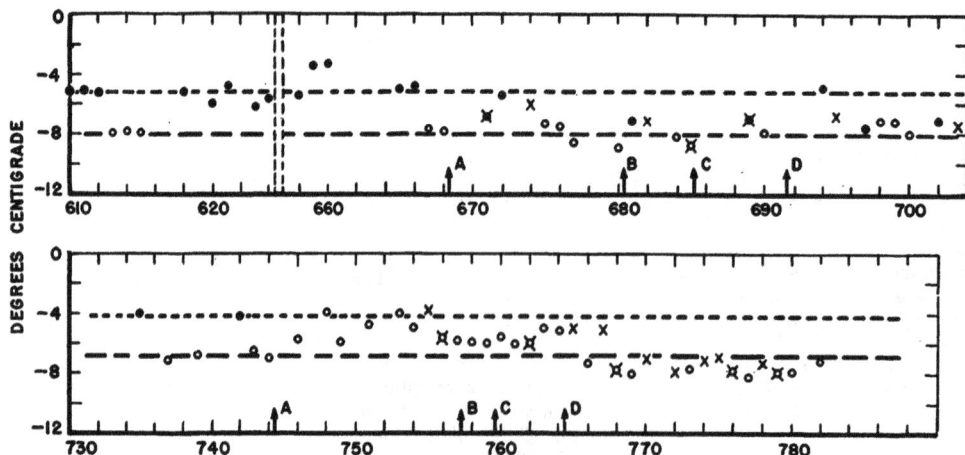

FIG. 30. Effect of solutes on freezing by rubbing: I. Glass on glass; only clean glass in the water. Ordinates: temperature at which freezing occurred (spontaneously, rings; on rubbing, dots); failed to occur on rubbing, crosses. A combination of ring and cross marks spontaneous freezing although rubbing at a fraction of a degree higher did not cause freezing. Abscissas: paragraph numbers in notebook. Initial volume of water, 35 ml. Arrows indicate the times at which aqueous solutions were added. Upper section: Added 0.12 ml NaOH at A and at B; 0.06 ml HCl at C and at D; after D, distinctly acid. Lower section: Added 0.09 ml chromic solution at A and at B, and 0.30 ml at C; 10.2 ml NaOH at D; after D, distinctly acid.

the distilled water to which it was added, but even a very small amount of any of them, sugar excepted, sufficed to prevent the initiation of freezing by rubbing at any temperature above t_{sf}. Sugar produced no significant change in either temperature.

These results accord with the others in indicating (1) that the presence of dissolved substances has little or no effect upon t_{sf}, and (2) that there is no intimate relation between that freezing point and the temperature at which freezing may be induced by rubbing one substance on another; and in addition they indicate that (1) the presence of even a small amount of an electrolyte may greatly depress the limiting temperature at which freezing is induced by rubbing, even making such induction impossible, but (2) the presence of even a large amount of a non-electrolyte may be without effect upon that limiting temperature.

But it should be remembered that these conclusions, resting solely on work done in a very dusty room and, consequently, with waters having a comparatively high t_{sf}, should be accepted with reserve. Before positive conclusions can be validly drawn there must be many more observations made under excellent conditions as to the cleanliness of the air, and extended to a variety of waters varying widely in their several values of t_{sf}. And it is very desirable that the effects of other solutes, both electrolytic and non-electrolytic, be studied, including those yielding ions of valence greater than one. And in particular, attention should be given to the differences that exist in the tendencies of the several

solutes to become adsorbed upon the surfaces that are rubbed together.

III. MISCELLANEA

Herein are collected certain incidental observations, conclusions, and remarks, that may be worth recording although not at all essential to the main purpose of this investigation.

1. ADHESION OF ICE

During the course of the work on freezing, a question arose as to the tenacity with which ice formed in contact with such substances as paraffin and bitumen might adhere to them; and, in particular, whether that adhesion is markedly less than that to surfaces over which water freely spreads. What was wanted was a rough and ready estimate, applicable to practical cases; and unless the maximum adhesion that might be expected proved to be much lower than in the case of easily wetted surfaces there would be no point in continuing the investigation. Pairs of wooden cylinders, ½ inch in diameter by about ¾ inch in length, were used. In the center of one end of each cylinder was a screw eye; the face of the other end was normal to the axis of the cylinder and was coated with the substance under test. The two cylinders of a pair were placed in a wooden clamp lined with paraffined paper, with their axes in line and their coated faces a few millimeters apart. The space between those faces was filled with distilled

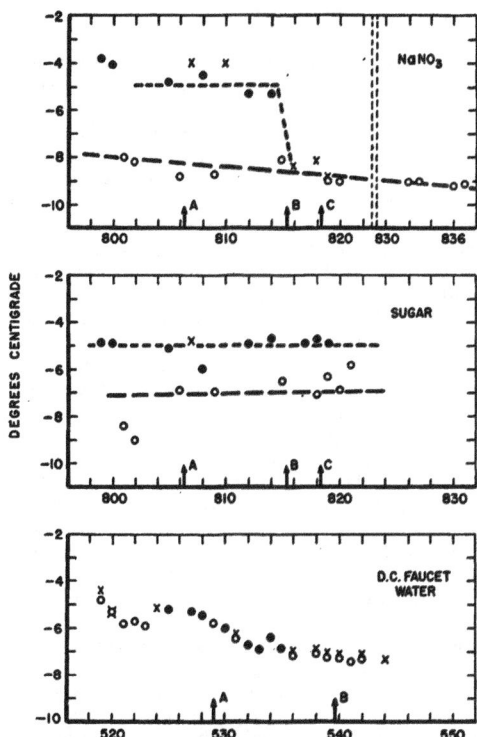

FIG. 31. Effect of solutes on freezing by rubbing: II. Glass on glass; only clean glass in the water. Ordinates: Temperature at which freezing occurred (spontaneously, rings; on rubbing, dots); failed to occur on rubbing, crosses. Combination of ring and cross marks spontaneous freezing soon after rubbing failed to initiate freezing. Abscissas: paragraph numbers in notebook. Initial volume of water, 35 ml. Arrows indicate the times at which aqueous solutions were added. Upper section: Distilled water. Solutions of NaNO$_3$: 0.22 ml weak solution at A; 0.64 and 0.86 ml of saturated solution at B and C, respectively. Middle section: Distilled water. Solutions of cane sugar: 0.43 ml weak solution at A; 0.86 ml solution 10 times as concentrated at B; 0.86 ml saturated solution at C. Lower section: D.C. faucet water. Strong solution of HCl: 0.30 ml at A; 0.33 ml at B.

water and the whole was placed in the ice-cube compartment of an electrical refrigerator until the water froze. The pair, now frozen together, was quickly removed from the clamp, hung from a spring balance, and subjected to an axial tension in a vertical line.

It was found that whether the cylinders were coated with Parowax or bitumen or were uncoated wet wood the frozen pair might withstand a pull of 15 lb., the limit of the balance. That corresponds to a tension of about 5 kg/cm², whereas the recorded values for the tensile strength of ice range from 2.4 to 25 kg/cm².

Whence it seems that in all three cases the adhesion may be of the order of the tensile strength of ice.

There were many failures, probably because the tests were made at laboratory temperatures, no cold box being at once available.

2. WETTING OF SURFACES

The idea that ice formed on paraffin might adhere to it only slightly arises from the rather common statement that water does not wet paraffin. Actually water does wet paraffin; but it does not spread over it, and one may perhaps say that paraffin is not readily wetted by water. But if the normal to the plane under-surface of a paraffin plate dipping into water be inclined to the vertical by less than the supplement of the angle of contact between water and paraffin, then water may be seen to rise against the under surface of the paraffin, as truly as it rises against a surface of clean glass. Furthermore, a small drop of water need rest on paraffin in ordinary air only a short time in order for it to become so firmly attached that it cannot be displaced without leaving a tail; and bits of paraffin can be readily lifted by touching them with a moistened finger tip. All of which shows that paraffin is, under usual conditions, wetted by water. Consequently, it is not surprising to find that ice frozen to it adheres strongly.

3. A BEAUTIFUL HABIT OF ICE

When freezing is induced in supercooled water by rubbing together in it two sticks soaked in oil—castor oil, Nujol, and kerosene were used—the ice takes on a radiating and curled appearance, quite suggestive of curled ostrich plumes. The radiating lines start from the region rubbed, but other centers quickly develop. The entire process is very rapid.

4. BURSTING OF WATER PIPES

In 1916 Dr. F. C. Brown [19] published the results of his investigations concerning the common observation that the "pipes carrying the hot water from the furnace to the kitchen and bath room burst from freezing more frequently than do the pipes carrying the cold."

He wrote:

In view of these experiments it has been concluded that the occluded air in ordinary tap water is responsible for the delay or absence of bursting of the pipes [carrying the cold water]. The air and accompanying impurities assist in furnishing nuclei of crystallization, so that the ordinary tap water begins to freeze at zero degree. At the same time the ice formed is more mobile, especially near the middle of the tube, so that until very low temperatures are reached the pressure is released along the middle of the tube by the flow of water and ice. In addition the air bubbles displacing water form cushions which relieve the pressure on the tube to a certain extent.

From this it has been quite widely inferred that the more frequent bursting of the pipes of the hot-water system arises from an absence of air in the water that had been heated, that absence depriving the water of the air-cushioning effect that would otherwise exist and that is effective in protecting the pipes of the cold-water system.

As will presently be shown, this inferred explanation is incorrect. Although it seems to be a fair inference from the quoted sentences, other portions of Dr. Brown's paper suggest that he attributed the more frequent bursting of the pipes of the hot-water system to a quite different proximate cause, and from that he reached the conclusion herein quoted by an acceptance of widely held erroneous beliefs.

His experiments were of this nature: Ordinary tap water was placed in each of a number of test tubes; half of those specimens were boiled, so as to expel the air; all were brought to room temperature; and then all were exposed to cold outdoor air. He found that the boiled water "invariably undercooled [supercooled] several degrees below zero before freezing commenced and . . . after crystallization began the temperature remained at zero until the entire mass was frozen. The ice was quite clear and solid." And that "the unboiled water always began freezing at zero and the ice was full of air bubbles and appeared quite slushy particularly near the central axis of the tube."

He further wrote: "The explanation then seemed rather to involve the super-cooling of the boiled water and the more solid ice formed in it. Both of these are largely dependent on the diminished amount of air in the boiled water."

The last statement is incorrect. It arose from an acceptance of two widely held false beliefs: (1) That the presence of dissolved air is inimical to supercooling, and (2) that the water in the hot-water pipes is essentially air-free.

That the water in the hot-water system under consideration contains much air (at least in Washington, D. C.) may be readily seen by drawing a beaker of hot water from the faucet. It will be seen that the water at first contains a dense cloud of minute air bubbles derived from the water. The phenomenon is particularly striking in cold weather, when the water taken into the water system contains much more air than when the weather is hot. It is not surprising that air should remain in the hot water; the water is heated in a closed system under considerable pressure.

That dissolved air is not inimical to supercooling, and that the increased supercooling subsequent to a heating of the water is not due to a removal of the air, have been shown by experiments herein reported. For example, in figure 4 it will be seen that t_{sf} for sealed specimen $P8$, containing air at atmospheric pressure, was $-6°C$, and after a 2-hour heating in boiling water it was $-14°C$. Furthermore, as early as 1916, Walton and Braun [90] had reported that aqueous solutions of O_2 and of CO_2 can be readily supercooled. The increased supercooling does not arise from an ex-

pelling of the air, but probably from a partial solution of the motes.

The correct explanation of the phenomenon is to be sought in the greater supercooling caused by the heating.

As the two systems of pipes, generally adjacent, are simultaneously cooled, the water in the cold-water system will begin to freeze when the temperature shall have fallen but little below 0°C. At that instant there is sudden freezing throughout the volume of supercooled water. That burst of freezing ceases as soon as the temperature has risen to 0°C, and results in the formation of only a little ice. Thereafter, and until the water shall have become frozen completely across the entire cross-section of the pipe, the water and ice at the axis of the pipe will remain at °C; freezing will proceed gradually from the wall inward; and water will flow along the axis of the pipe to the reservoir in amounts to compensate for the expansion that results from the freezing. The pressure on the pipe will remain essentially unchanged throughout that period, and the excess of the air rejected during freezing above that required to saturate the water at the pressure in the system will appear as bubbles, mainly adjacent to the advancing boundary of ice and water. However, after the water has frozen so as to block the pipe at any section, a continuation of freezing between the block and a dead end will result in an increase in pressure, and perhaps in a bursting of the pipe. That cause of bursting is equally effective in the hot-water system.

During the second stage, after the pipe has become blocked, any formation of air bubbles would result in an increase in the pressure, since the density of air-saturated water differs but little from that of air-free water. It seems unlikely that any bubbles would form during this stage.

On the other hand, the water in the hot-water system will not begin to freeze until it shall have cooled considerably below 0°C; and the length of pipe then filled with supercooled water will, in general, exceed that so filled in the cold-water system when freezing began in it. As in the preceding case, the beginning of freezing is sudden, and on account of the marked supercooling, a relatively large amount of ice will form quickly throughout the volume of supercooled water. Even at the axis of the pipe the needles of ice may interpenetrate so closely as to form with the water a pasty, or an almost rigid mass, extending some distance along the pipe. Thereafter, freezing will proceed gradually from the wall inward, the temperature at the axis remaining zero until all the water in that section of the pipe shall have become frozen. But in contrast with the preceding case, the long, pasty or rigid mass formed in the pipe during the initial sudden freezing greatly impedes, and may completely block, the back flow of water to the reservoir. Then the pressure rises, the axial mass becomes more and more solidified, and the pipe may burst before all the water in that section of it shall have become completely frozen. Therein lies the essential explanation of the difference in the bursting tendency of the two systems of piping.

It is very regrettable that Dr. Brown's experiments, revealing as they did the correct explanation of the phenomenon, have been so seriously misinterpreted.

5. QUIET BOILING

At the beginning of this work great annoyance was caused by the severe bumping of the water in the boiler used for steaming the bulbs. Not only did the commonly advised expedients—such as the placing of beads, of pointed bits of platinum, etc. in the water—prove ineffective, but a little consideration made it evident that no help should be expected from the introduction of any well-wetted solid, whether pointed, rounded, or flat; for in all such cases both the initial volume of gas (steam) and the initial radius of curvature of the liquid-gas interface are vanishingly small, and the interface is everywhere concave toward the gas phase, thus increasing its pressure.

If, however, one could find a substance with which water has a large contact angle—the larger the better, and certainly larger than 90°—then the initial radius of curvature of the liquid-gas interface might be finite and consequently the capillary pressure need not be excessive even though the initial volume of the gas be small and the interface be concave toward it. And if the substance were also porous, there would be in each pore a volume of gas partially bounded by a liquid-gas interface convex towards it. In each case, evaporation would be facilitated. No suitable substance of that kind was found.

Substances that take up a large amount of air (charcoal, bitumen, metallic soaps, etc.) are exceedingly efficient when first placed in water, but as the air is boiled away their efficacy decreases, and after the air has been completely replaced by steam they become entirely inert if the temperature falls for a time below the boiling point. On exposure to the air they recover their efficacy, and some will recover while still immersed in the water if that is allowed to cool and to stand over night in contact with the air. They seem to recharge themselves with air derived from that which dissolves in the water.

No such substance suitable for the purpose in mind was found.

Another device frequently used for the purpose, when the boiling is not to be very rapid, consists of a length of small bore or capillary glass tubing closed at its upper end and supported, or stood, open-end down in the liquid. The air in the tube furnishes at its lower end an air-liquid surface at which vaporization can readily occur, the excess air-vapor mixture escaping from the lower end of the tube. That device works excellently. The air, however, is gradually replaced by vapor, and if the temperature then falls below the boiling point, the tube fills with water and becomes inert until it has been removed and again air-filled. That is a distinct disadvantage.

The device finally adopted is based on this same principle, combined with a utilization of the following well-known phenomenon. If the end of a clean glass rod be drawn slowly along the bottom of a clean beaker containing superheated water, and being continuously heated from below, a stream of bubbles will, in general, rise from the region of contact of the moving rod with the beaker, even though the superheating be slight. Consequently, if the open end of the tube considered in the preceding paragraph be moved along the bottom of the vessel, continually maintaining contact with it, the tube may become recharged by some of the rising bubbles, provided that the diameter of the tube is not too small. In fact, it does. Furthermore, even though it does not move over the bottom of the vessel it will usually become charged by bubbles produced during the preliminary gentle bumpings caused by the superheating of the thin layer of water between the edge of the tube and the bottom of the vessel.

A very convenient device for securing the desired conditions may be constructed as follows. One end of a short section of glass tubing is ground flat and approximately perpendicular to the axis of the tube, and one or more shallow V's are cut in the tube's walls at that end.

This stands on the bottom of the vessel and into its top is passed a loosely fitting vertical glass rod, the space between the rod and the wall of the tube being so narrow that steam will not pass when its pressure is such that it escapes freely through the V-notches in the bottom of the tube. The rod serves to keep the tube upright, to regulate the length of the vapor space in it, and to slide the tube slightly along the bottom of the vessel; it may be supported by a clamp stand, by being passed through a rubber stopper closing a flask, by a spider resting on top of a breaker, or by any other convenient means. As the boiling point is approached, the section of tubing is now and again caused to slide slightly along the bottom of the vessel, either by tapping the rod or by jarring the vessel; and presently that will result in an emission of bubbles, many of which will escape, but some of which will be caught under the rod in the tube. The first of those will probably collapse at once, but after the temperature of the water, the bottom of the rod, and the tube have reached the boiling point, the caught bubbles will persist and grow, and a lively stream of bubbles will issue from the V's in the bottom edge of the tube. Even though the tube be not slid along the bottom of the vessel it will presently become charged by gentle bumping from under its edge.

It is desirable that the bottom of the rod be within a few millimeters of the bottom of the vessel, so as to insure quick charging.

The dimensions of the parts are of minor importance. I have commonly used tubing 5 or 6 mm in internal diameter and 15 to 30 mm long, and rods about 1 mm smaller in diameter than the lumen of the tube. The V's, generally two in number, were about 2 mm high.

In certain cases, as in distillation, it is desirable to have the entire device inside the boiler and it is inconvenient to support the rod independently. In such cases the section of tubing may be fused to the rod, the length of the latter being so adjusted that, when the cup rests on the bottom, the top of the rod will project well into the neck of the boiler, and to the rod two suitable glass spiders may be fused, one at its top and the other as much lower as is consistent with its remaining in the neck however much the inclination of the rod may be changed. That keeps the terminal cup approximately vertical, keeps a portion of its rim in contact with the bottom of the vessel, and enables one, by jarring the vessel, to cause a slight sliding. That worked quite satisfactorily, but is not to be preferred to the loose tube device. The cup I used was 15 mm long; a much shorter one would prime more quickly.

C. THEORETICAL CONSIDERATIONS

I. INTRODUCTION

Since long before this work was begun, it had been known that the spontaneous freezing of a supercooled melt begins only at certain discrete points, the number of those singular points being extremely small, almost vanishingly small, as compared with the number of molecules in the melt [30, 78].

At those singular points, either there are foreign particles or the local state of the melt at the instant just before the freezing differs from that elsewhere. There seems to be no other alternative. At this time, one need not consider whether the mentioned local difference in state arose immediately before the initiation of the freezing process, or whether it had preexisted for a significant interval of time. But if it had preexisted, then some special situation affecting it must have arisen at the instant that freezing began.

Those initial local differences in the state of the melt, or those foreign particles, as the case may be, which characterize the singular points at which freezing begins, have frequently been called "nuclei," suggestive of the fact that the freezing starts at those points and spreads progressively from them.

Obviously, the spreading crystallization may entirely surround the singular point from which it started. In such case, the term "nucleus" becomes still more appropriate. Or one might then call it the "kernel" of the crystallization; or since the crystallization originates there, it might be called the "germ" of the crystallization. All these terms have been used. Each of the terms "nucleus" and "kernel" is quite appropriate when

the singular point is characterized by a foreign particle which becomes surrounded by the spreading crystallization, but neither seems appropriate to the case in which the singular point is characterized merely by a peculiar condition of the melt which favors the initiation of crystallization. And "germ" would seem to be much more appropriate to the initial step in the freezing—the initial, persisting, structural aggregation of molecules that presently becomes a crystal—than it is to either a foreign particle or a small region in which the state of the melt is peculiar, but still definitely that of a liquid.

Furthermore, the terms "nuclei," "Keim" (germ), and "Kern" (kernel) have each been used to denote not merely the singular points at which crystallization begins, but also the structures that are gradually built up at, and spread from, those points [43, 78, 79]. And each of the terms "nucleus," "kernel" (Kern), and "germ" (Keim) has been, and is, used with different restrictions by different authors [7]. The several terms and modifications of them—"Unterkeim" (subgerm) [47, 76], "Ultrakeim" (ultragerm, sometimes to denote structures smaller than the germ, sometimes to denote those that are larger) [43], "ultra nucleus" [6], "germ nucleus" [6], "grains" [6, 7], and others—are to be found in papers dealing with the freezing of supercooled melts or with the crystallization of supersaturated solutions.

The existence of such a confused and not always appropriate nomenclature indicates the unsettled state of the subject, and greatly increases the difficulties of the student. Such confusion should not be perpetuated. But a mere redefining of primary terms that have been so abused will merely add to the confusion. What is needed is a new nomenclature that breaks with the old so clearly that none can fail to see that it is something different. Each term should be clearly defined, and in some obvious way it should be appropriate to each of the situations in which it is to be used, and it should have as wide a coverage as possible, restricted coverage being indicated by the use of adjectives, not prefixes. Cases where prefixes may seem to be desirable had best be covered by independent terms, the primary term being so defined as to limit its coverage accordingly. The following terms seem to differ sufficiently from those previously used, to be closely appropriate to the ways in which they will be used, and to meet all the present requirements. They will be used throughout the rest of this report.

Singularities: Each of the singular points at which freezing is spontaneously initiated in a supercooled melt will be called a singularity. Two distinct types of singularities will be recognized. They will be distinguished by the adjectives *homogeneous* and *heterogeneous.*

Homogeneous singularities: A homogeneous singularity is a singularity characterized solely by a local difference in the state of the melt, that difference arising fortuitously and solely from the thermal agitation of the molecules.

Heterogeneous singularities: A heterogeneous singularity is a singularity characterized by the presence of something foreign to the melt. It would not be a singularity but for the presence of that something. Furthermore, this definition does not imply that the presence of an interface with a foreign substance will necessarily result in a singularity in the sense in which that term has been here defined. It may also be remarked that although the presence of an interface between the melt and a foreign substance may modify the state of the adjacent melt such modification should never be confused with a homogeneous singularity.

Embryos of crystals, or more briefly, *Embryos:* [1] An embryo of a crystal is any structural aggregation of the molecules of the melt that maintains its identity as an individual, distinct from the ambient melt, for an interval that is long as compared with the mean time between consecutive molecular collisions, and that will under favorable conditions become an ordinary macroscopic crystal without ever having lost its identity. The retention of identity as an individual does not preclude a limited mutual transfer of molecules between it and the ambient melt, whether such transfer results in a change in size or no.

Nothing need be said about the intimate structure of the embryo; it will probably change as the embryo changes in size and age, approaching that of the crystal as the embryo approaches the crystal state. As will presently be seen, the stability of an embryo varies with its size; so that an embryo of a given kind can at a given temperature be in equilibrium with its melt only if it is of a particular size. That will be called its *critical* size, and at times an embryo of that size may be called the *critical embryo.* Embryos larger than the critical will *mature* into macroscopic crystals; those smaller than the critical will wane (decay) to evanescence. Hence, at any given temperature the initiation of a *viable* embryo involves the origination *per saltum* of an embryo of at least the critical size.

Matured embryos: An embryo that is growing towards the crystal state may be said to be maturing, and one that has apparently or presumptively reached that state may be called a matured embryo, rather than a crystal, when one wishes to indicate that the crystal state may not yet have been fully attained.

II. EMBRYOS

1. INTRODUCTION

One studying the spontaneous freezing of a supercooled melt is interested in the singularities in that melt only (1) as places at which the conditions are such

that the process of freezing can be spontaneously initiated, and (2) to the extent to which those conditions throw light upon the manner in which that process is initiated. Once the process is started, his interest is concentrated on the growing region of more or less definite structure, formed from the substance of the melt and destined to become presently an ordinary macroscopic crystal; that is, on the growing embryo.

Questions relating to the number and types of singularities that are present and to the manners in which the embryos are initiated in each of the several cases, will be considered in a later section devoted to the several theories of freezing.

Here will be discussed the properties, stability, growth, and decay of an embryo.

It should be remembered that the definition of an embryo of a crystal says nothing about the actual structure of the embryo, providing indeed for possible changes in that structure as the embryo waxes or wanes. But it does require that the embryo shall be a durable entity, persisting as an individual throughout numerous molecular encounters, until, if waning, it presently becomes so small that it vanishes suddenly, being broken up by a violent molecular encounter. A mere transient geometrical arrangement formed by molecules while passing through the positions then instantaneously occupied by them does not constitute an embryo, however closely that transient arrangement of molecules may approach that which characterizes an embryo; not even though the x-ray diffraction pattern formed by that transient arrangement should be essentially the same as that formed by an embryo—as perhaps it may, the period of the rays being so vastly smaller than the time between consecutive molecular encounters.

2. STABILITY AND GROWTH

a. Simple Embryo

An embryo suspended in its melt is subjected to a continuous molecular bombardment from without; and its own molecules are constantly endeavoring to escape. Each of these influences favors its disintegration. On the other hand, its forces of molecular attraction tend to restrain the escape of its own molecules and to capture the more slowly moving of the impinging ones. These favor its growth. If the latter overbalance the former, the embryo grows; if the contrary, it decays. If the two balance, its size remains unchanged. But even in this case of equilibrium, there is a constant interchange of molecules between the embryo and the melt surrounding it. It is constantly losing its more swiftly moving molecules and as constantly capturing the melt's more slowly moving ones.

From general considerations (classical theory of capillarity, increase of vapor pressure with the convexity of the surface, that an impinging molecule can be applied more efficiently for the removal of surface

---

[1] Only after this section had been written did I find that P. Pawlow [64] had used the term "embryo," but not in the wide sense in which it is used here.

molecules if the surface is highly convex than if it is flatter, that very small bodies may be expected to hold on to their surface molecules less strongly than larger ones, etc.) it seems likely, and may be assumed, that the smaller the embryo, the more easily will it lose molecules and the more difficult will it be for it to capture impinging ones. Consequently, the smaller the embryo, the lower the temperature at which it will be in equilibrium with its melt. Furthermore, the equilibrium will be essentially unstable; if the size is ever so slightly smaller, other conditions remaining the same, the embryo will decay to evanescence; if it is ever so slightly larger it will mature to a crystal. That size which is in equilibrium with the melt has just been defined as the "critical" size of the embryo under the existing conditions. From all of which it is obvious that in the initiation of embryos they must *per saltum* attain at least their critical size. Since that increases as the temperature rises, it is obvious that the initiation of embryos is far more difficult at high temperatures than at low, and may be actually impossible except under special conditions that antecedently assure the existence of a large group of closely neighboring molecules having small relative velocities.

In the preceding, it has been assumed that only a single type of molecule is present—that of the pure melt. If there are other types of molecules, some of them may tend to accumulate on the surface of the embryo; and under certain conditions this may result in a stability that would not otherwise exist. This should be remembered, but need not be further considered now.

Those who wish a more complete and mathematical study of the stability of embryos are referred to Gibbs' classical article: "On the equilibrium of heterogeneous substances" [38]. Detailed treatments of the following problems pertinent to the present study will there be found: (1) Liquid drops in their own vapor (p. 219); (2) the initiation of a new phase at a point in the interior of a preexistent homogeneous phase (p. 252); and (3) surfaces of discontinuity between solids and fluids (p. 314). The treatments are basically very similar. In every case, there is defined a certain hypothetical "surface of tension" S surrounding the interior phase. To that surface is assigned such an amount of energy that the sum of it and the energies of the two phases, each supposed to extend without change to the mathematical surface S, is equal to the actual energy of the entire system. The quotients of the energy assigned to S divided by the area of S he denotes by σ, and calls the "surface tension" of S. Thus he gets a similarity of treatment. But one should notice that he carefully states that σ "cannot properly be regarded as expressing the tension of the [fluid-solid] surface" (p. 315).

Our present interest centers in the second of those three problems—the initiation of an embryo—and it is very important to notice how he defines the surface S when the fault in homogeneity which it surrounds may be so small that no part of it can fairly be assumed to be homogeneous (p. 253). He shows that at any given temperature and pressure there is a unique size of S, regarded as spherical, that is in equilibrium with the surrounding phase; smaller ones decay; and were an embryo of the equilibrium size formed by an external agency, its equilibrium would be "unstable, and with the least excess in size, the interior mass would tend to increase without limit except that depending on the magnitude of the exterior mass" (p. 256). He shows that if the new phase can form spontaneously—if there be a limit at which the initial phase becomes unstable— "it must be at or beyond the limit at which σ vanishes" and "if the fault in homogeneity of the mass vanishes at the same time (it evidently cannot vanish sooner), the [initial] phase becomes unstable at this limit. But if the fault . . . does not . . . [then] vanish—and no sufficient reason appears why this should not be considered as the general case—. . . [the simultaneous vanishing of r, w, and σ] is not enough to make the [initial] phase unstable" (p. 256). "The real stability of a phase extends in general beyond that limit . . . at which the phase can exist in contact with another at a plane surface" (p. 257).

All of which confirm the conclusions already drawn from molecular considerations.

But how rapidly does an embryo grow? and how does its rate of growth depend upon its size and temperature? and how do all of these depend upon the nature of the substance? To these questions, considerations of stability, such as those of Gibbs, give no answer. Some insight into the problems so posed may, however, be obtained by a consideration of a simple, ideal case.

Consider a spherical embryo of radius r at a uniform temperature t_0 immersed in the center of a large volume of melt at temperature t at points far from the embryo. Let ρ, l, and v be, respectively, the effective density, latent heat, and volume of the embryo, and c be the thermal conductivity of the melt. Assume that ρ, l, and c are independent of both t and r. If l is finite, t_0 must differ from t when r is changing.

The heat freed at the surface of the embryo during the interval $d\tau$ will be

$$l\rho(dv/d\tau)d\tau, \qquad (1)$$

that conducted away during the same interval will be

$$4\pi cr(t_0 - t)d\tau, \qquad (2)$$

the resistance between a small sphere of radius r and a concentric one of infinite radius being $1/4\pi cr$ [55].

Hence, for a steady state

$$dv/d\tau = (4\pi c/l\rho)(t_0 - t)r. \qquad (3)$$

On replacing dv by its equivalent $4\pi r^2 dr$, one finds

$$dr/d\tau = (c/l\rho)(t_0 - t)/r, \qquad (4)$$

when $t = t_0$, $dr/d\tau = 0$; hence t_0 is the temperature of the melt at which the critical embryo has the radius r.

The effect of the nature of the substance is determined by the values of c, l, and ρ, and by the relation connecting t_0 and r.

If the embryo is larger than the critical one corresponding to t, then $(t_0 - t)$ is positive, and the embryo will grow. But as the embryo grows, t_0 will increase; and it may be that presently $(t_0 - t)/r$, and consequently $dr/d\tau$ also, may become constant.

If the embryo is smaller than the critical one corresponding to t, then $(t_0 - t)$ is negative, the embryo will decay, and as r decreases so will t_0, and $(t - t_0)$ will increase. Consequently, $(t - t_0)/r$ and the rate of decay will increase as r decreases. And if l decreases with r, as one would expect it to do when r is very small, that will enhance the rate of decay. A simple embryo that is smaller than the critical one will decrease continuously at an ever increasing rate until it vanishes.

b. Complex Embryo

If there be in the melt a foreign particle that adsorbs molecules of the melt and becomes enveloped by an embryo, the combination may be regarded as a *complex embryo*. For simplicity, suppose that the particle is spherical. If the thickness of the covering of the particle is less than the range of the particle's force of adsorption, the complex embryo will attract and hold molecules of the melt more strongly than will a simple embryo of the same size. That is, its effective size will exceed its actual size; and that excess will increase as the thickness of the covering decreases. But if the thickness exceeds the range of the force of adsorption, then the complex embryo will act as a simple one of the same size.

The variation of the effective radius r_e of the complex embryo with the actual radius r is somewhat as shown diagrammatically in figure 32.[2] If one uses subscripts to denote the point to which a particular r refers (e.g. if r_B indicates the actual radius of the complex embryo that corresponds to the point B in the figure), then r_D is the radius of the particle, and $r_A - r_D$ is the range of the force of adsorption. The point C is where the tangent to the curve is horizontal. Throughout the portion from A through B, $r_e = r$; but from A through C to D, $r_e > r$. Within the range defined by BCD, there are pairs of complex embryos having the same effective radius; e.g. those having radii r_P and $r_{P'}$. If the complex embryo is in a melt at such a temperature t that the radius of the critical

---

[2] Obviously the variations in r at any given point of the surface are discontinuous, each step being determined by the molecular diameter. But, while molecules are being added at one place, none will be added at others, and at still others molecules will be removed. Consequently, the mean radius may, for present purposes, be regarded as capable of a continuous variation.

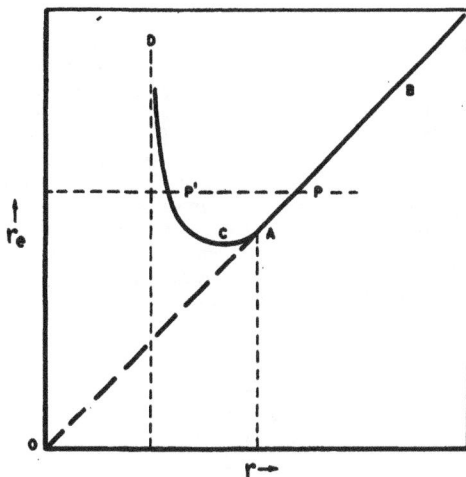

Fig. 32. Effective radius of complex embryo. Schematic only. Ordinates: effective radius r_e of the complex embryo. Abscissas: actual radius of the embryo. r_D is actual radius of spherical mote forming the core; $r_A - r_D$ is range of adsorptive forces; C is minimum value of r_e. From A through B, $r_e = r$.

embryo is r_P, then, if the effective radius (r_e) of the complex embryo exceeds r_P, r will grow continuously. But if r_e is less than r_P, r will decrease, through C to P', at which point r_e is again equal to the radius of the critical embryo, and there its decrease ceases. A complex embryo having a radius that is less than r_0, is thoroughly stable if its effective radius equals that of the critical embryo. A complex embryo cannot have an effective radius that is less than that corresponding to the point C.

If the force of adsorption is strong and the particle is not too small, it may be that the complex embryo on the CD portion of the curve will be stable up to a temperature t' well above the melting point of a crystal of the solute. In such a case the embryo will not be destroyed by heating to any temperature lower than t'. If, after such heating, the melt containing it be cooled below the melting point of the substance to a temperature t for which the corresponding critical embryo has a radius that is at least as great as the value of r_e corresponding to C, then the complex embryo will grow in actual size, and decrease in effective size, until r_e shall have become equal to that of the critical embryo. Its growth will then cease. But if t be such that the radius of the critical embryo is smaller than the value of r_e corresponding to C, then the growth of the complex embryo will continue beyond C, through A and B, until the entire volume of the melt is frozen.

If the melt be heated above the temperature t', the complex embryo will be totally destroyed, will be reduced to merely the particle with its adsorbed layer of the melt. Nothing that has been said regarding embryos gives any clue as to whether a recooling of the melt will result in a re-forming of the complex embryo, nor as to the temperature at which such feasible re-forming might be expected. However, if the melt were, by any means, caused to freeze, it would seem that the particle must then become enclosed by the frozen substance in exactly the same manner as if the freezing had been initiated by the complex embryo; and in that case a remelting of the substance at a temperature below t' would leave the particle as a core to a complex embryo of the kind already studied. The effect of a heating would be annulled by a subsequent freezing.

Consequently such complex embryos cannot account for the permanence of the effect produced by preheating the melt, but they may account for an occasional initial t_{sf} being lower than the immediately following ones, for those effects commonly described as arising from the melt's "remembrance" of its immediately preceding state, and for the reported greater ease with which freezing can be initiated by rubbing those portions of the walls of the container that had been in contact with the frozen substance immediately preceding the melting than by rubbing those portions that had not, but which were nevertheless wet with the melt at the time of rubbing. If such effects arise in this manner then it should be possible to eliminate them by heating the melt to a sufficiently high temperature; and it may well be that the duration of heating is of importance, especially at the lower temperatures. Time is required to complete the change.

Such complex embryos seem to be much the same as what earlier workers have described as particles carrying a crystalline adsorbate [67], and have invoked for the purpose of explaining the effect of preheating the melt; for which, however, they seem to be unsuited.

Whether such complex embryos actually exist and behave in the manner here described cannot be determined satisfactorily from the observations herein reported on the freezing of water, but there seems to be no reason for doubting that conditions may arise under which they would be formed and would probably behave in the manner indicated.

3. STRUCTURE

One naturally asks: What is the intimate structure of an embryo? How does that structure change as the embryo grows? At what stage does one regard the embryo as fully matured—as an ordinary macroscopic crystal?

Taking the last question first, an embryo is a fully matured crystal when it is in internal equilibrium and in equilibrium with its melt at the same temperature as is an ordinary crystal of the same size.

As regards the other two questions, the only safe answer is that we know nothing about the embryo's structure when it is very small, but near maturity, its structure must be very nearly that of the crystal lattice corresponding to the macroscopic crystal. Whatever may be its initial structure, there is a progressive, but not necessarily continuous, change toward the structure characterizing the crystal. Experiment alone can determine how the changes actually occur.

4. OUTWARD FORM

The embryo's outward form, like that of a snowflake, may perhaps vary widely from embryo to embryo. A few observations of the outward forms of matured, or nearly matured, embryos, soon after they become visible, have been reported. Ostwald [61] stated that Link [51] observed in 1839 that the earliest microscopically visible precipitate from a solution is in the form of liquid drops, which coalesce and presently attain a solid crystalline state; and that between the drop and the final crystal there are manifold intermediate forms, called by Vogelsang [87] crystallites. Frankenheim [32] accepts most of Link's observations, but criticizes his interpretation of them. Tammann, who studied many substances, has reported [78, 79, 85] that the early structures are usually spherulites formed of many fine needles radiating from a point; in some cases one or more of the needles are much coarser and longer than the others. Sometimes a star-like structure of small columns radiating from a point is formed, and only very rarely are isolated regular crystals observed. H. T. Barnes [11, 12] has stated that, at its first appearance, ice has the form of circular disks. That is confirmed by Altberg [1, 2, 3, 4] who, by working with water only very slightly below 0°C and by keeping the growing structure floating freely within it, has followed the structure's growth. He has stated that the fully developed embryos of water are regular discs of transparent ice a few millimeters in diameter, perfectly round and with smooth surfaces. They are very hard to see. These disks become changed into regular hexagonal plates and then into six-rayed stars, still with perfectly smooth faces. Projecting patterns form on the stars, transforming them into shapes suggestive of snowflakes, which may attain a diameter of over 2 cm.

I have not tried to verify these observations. It would seem that they all refer to the early stages of ordinary macroscopic crystals, not to what has here been defined as an embryo; but only a careful study of their thermal equilibrium with water can decide that question. It may be that their normal melting points are slightly below 0°C. And it may be that the processes of growth and decay differ in such a way that the limiting size at which a growing structure comes to equilibrium with water at 0°C is larger than that at which the decaying structure ceases to be in such equilibrium. That is, it may be more difficult to arrange the molecules exactly in the ultimate crystal lattice, than it is to maintain them there; so that in the first case a larger volume is required than in the second. It's a matter to be decided by experiments yet to be made.

III. RECOGNIZED THEORIES OF FREEZING

1. INTRODUCTION

Corresponding to each of the two kinds of singularity —homogeneous and heterogeneous—there was when this work was begun, and there is today, a theory of freezing that has numerous supporters. The one stressing the importance of homogeneous singularities will be called the *Homogeneous Theory;* the other, the *Heterogeneous Theory.*

Most upholders of each of these theories acknowledge the possible efficacy of the other type of singularity in certain cases. But advocates of the homogeneous theory regard all effects arising from heterogeneous singularities as disturbances superimposed on the main phenomena; and those of the heterogeneous theory regard effects arising from homogeneous singularities as negligible in many cases that they have investigated, although of possible importance in certain others.

The difference, though apparently one of emphasis, is actually profound. Upholders of the homogeneous theory, as it has been, and is currently, expounded, cannot accept any suggestion that it may in certain cases be invalid. On the other hand, upholders of the heterogeneous theory, basing their claims on experimental evidence, cannot accept mere unsupported guesses as to how the data might conceivably be explained in another way. Every experimenter acquires much information that cannot be satisfactorily presented in a written paper, but which is invaluable to him in interpreting his data, and in assessing the validity of proposed interpretations of them.

In the following pages each of these two theories will be discussed. Then, pertinent experimental data will be compared with the demands of each theory. It will be found that such a comparison indicates that a new approach must be made to the problem. The outline of such an approach will then be presented.

An attempt has been made to choose the wording throughout the text in such a way as to be applicable to the freezing—crystallization—of a melt, but with slight obvious modifications what is written will be equally applicable to the crystallization of solutes from their solutions; and in any appeal to experiment no distinction will be made between observations on melts and those on solutions, it being generally admitted that the main features of the two processes are essentially the same. One should, however, remember that in the crystallization of a solute factors may enter that are of little or no importance in that of a melt. For example, in the case of a solute diffusion plays an essential role, and, as pointed out by Graham [42] over a century ago, the effect of dissolved gases upon the solubility needs to be considered.

2. HOMOGENEOUS THEORY

a. The Theory

The homogeneous theory is the first and only clearcut theory of freezing that has been proposed. It was suggested by de Coppet [21, 22, 23] in 1872, developed by him in 1875, and reaffirmed by him in 1907. It rests on opinions regarding the kinetic theory of matter, and has had many other adherents, among whom may be mentioned G. Tammann, S. W. Young, W. E. Burke, P. Othmer, W. J. Jones, J. R. Partington, K. Schaum, H. T. Barnes, S. S. Kistler, J. Frenkel.

The theory states that freezing is initiated at homogeneous singularities—at those places where such suitable fortuitous encounters of molecules of the melt occur that the colliding molecules can become so bonded together as to form a persistent unit (an embryo, in our terminology) which then, if the temperature is right, grows by progressive accretions, possibly changing in internal structure, and presently becomes a visible crystal of the substance.

Adherents to the theory tacitly assume that the nature of the encounter of molecules at a given point and time is entirely fortuitous, following strictly the laws of chance with respect to every factor involved—speed, orientation, density of packing, etc. They also assume that each one of those factors may, and in the course of time will, take, as a result of chance interactions, every conceivable real value.

Whence it follows that they tacitly assume that encounters suitable for the formation of any given size of embryo may, and in time will, occur at any given point in the melt, whatever the temperature may be, provided only that the viscosity of the melt be not so great as to prevent the required motions of the molecules. At temperatures above the melting point of the substance, all the embryos so formed are evanescent; at lower temperatures some of them are viable. Hence, however small the supercooling may be, freezing will presently occur; a supercooled melt is an essentially unstable system, is in labile equilibrium.

Furthermore, since the minimum size of a viable embryo increases with the temperature, the higher the temperature the greater must be the multiplicity of the collision that results in such an embryo. But obviously the chance of a multiple collision occurring decreases with its multiplicity. Hence, the higher the temperature the smaller the probability that a viable embryo can be so formed.

Since the ease with which colliding molecules can so arrange and bind themselves as to form an embryo is obviously greater when their relative velocities are low, an embryo of given size can be formed the more easily —the probability of its being formed is the greater—the lower the temperature of the melt, assuming that the viscosity is not too great.

Likewise, the chance of the desired embryo being formed is greater in those limited regions in which, owing to fluctuations, the molecular speeds are lower, and the orientations and packing are more favorable, than they are in the more normal surrounding melt.

The theory as just outlined demands certain phenomena susceptible of experimental study.

b. Phenomena Demanded by this Theory

Some of the phenomena more obviously demanded by the homogeneous theory are listed below. In a later section, these will be compared with pertinent experimental data.

A cautionary remark seems necessary here. Any actual specimen of melt contains an immense number of molecules. Does that mean that at any given instant in its history every molecular situation that arises at any time in an infinite volume of the melt, subjected to the same conditions, is to be found somewhere in the specimen? If so, all such specimens of the substance must yield the same result, and that same result must be found test after test. Furthermore, such specimens cannot be supercooled. As is well known, different specimens of the same substance seldom give the same result, and specimens can usually be supercooled. Whence it would seem that, in general, actual specimens behave in such a way that they cannot be regarded as of infinite size.

On the other hand, if the specimen cannot be regarded as of infinite size, then its behavior should be expected to vary from test to test, and that of one portion of its volume should be expected to differ from that of another, and no one portion should behave in the same way test after test. In such a case, repeatable results can be obtained only by averaging the results of a large number of independent observations taken under the same fixed conditions.

In the following list, the specimen is always assumed to be distinctly finite in size, unless the contrary is clearly stated.

1. A specimen should not freeze spontaneously at a fixed characteristic temperature.

2. If a large group of nominally identical specimens of a melt, all at the same temperature, slightly above the melting point of the substance, be suddenly immersed in a bath at a low temperature, then the number of specimens that freeze during the interval between τ and $(\tau + d\tau)$ should vary with τ in accordance with the law of chance, τ being the time that has elapsed since the instant that the group was placed in the cold bath. Furthermore, that variation will be the same for each of a series of successive repetitions of the test.

If the values of τ at which any given specimen freezes be observed test after test, and the number of times that it freezes between τ and $(\tau + d\tau)$ be plotted against τ, the resulting curve should be similar to that for the group.

3. As the constant temperature of the cold bath is progressively lowered by steps, the *average* time required for a given specimen to freeze should continuously decrease until the viscosity shall have become so great as to impede significantly the motions of the molecules. After that, a further lowering should generally cause an increase in the average time.

4. If two nominally identical specimens, differing only in the volume of the melt, be treated as just described, then *on the average* the one of larger volume should freeze before the other.

5. In any given supercooled specimen kept at a constant temperature, a singularity will occur from time to time. The number of such singularities that occur per second per unit volume of the melt should be a constant determined solely by the melt and the temperature, with which it may be expected to vary.

3. HETEROGENEOUS THEORY

a. The Theory

The heterogeneous theory of freezing rests solely on experimental data, and arose from the steady accumulation of observations that could not satisfactorily be accounted for on the homogeneous theory. In its present form it maintains that freezing is spontaneously initiated solely at heterogeneous singularities—at those places where the melt is in contact with another substance, either the walls or particles adhering thereto or foreign particles suspended in the melt.

Its adherents insist that this is surely true throughout the range of supercooling that is ordinarily studied.

A great advantage of this theory over the homogeneous one is that the first step in the formation of a viable embryo need not necessarily involve a *per saltum* formation of a structure of some size. It is conceivable that the molecules of the melt might be laid down upon the particle or wall, one by one, the attraction of the particle or wall furnishing the stability that would otherwise require a group of many molecules of the melt. Nevertheless, being completely at variance with the accepted form of the homogeneous theory, which appeared to have a sound theoretical basis, the heterogeneous theory was, and is, distinctly unpopular. Many, looking upon it as running counter to the kinetic theory of matter, regard it as subject to *a priori* condemnation. This situation is worsened by the absence of any consensus of opinion as to how the foreign particles act.

Under such conditions it is not surprising that only gradually and as evidence accumulated against the homogeneous theory did a few observers, now and then, advance towards the positive theory. It seems that Meyer and Pfaff [56, 57] were the first to announce clearly that no nucleus of freezing arises spontaneously in the interior of a homogeneous melt.

Most earlier observers were of the opinion that heterogeneous singularities are few in number, and do no more than superimpose minor irregularities upon effects arising from homogeneous singularities. Many, probably most, later observers are of the same opinion. Whenever those observers were forced to abandon the idea of minor irregularities, they sought a means for accommodating the homogeneous theory to the offending irregularities, and the prestige of that theory was such that mere plausibility was not infrequently deemed a sufficient ground for accepting the proposed accommodation, and thus discountenancing the heterogeneous theory. That was rendered easier by the paucity and the restricted scope of the observations commonly reported by an individual worker. All of which made it difficult for a heterogeneous theory to obtain a start.

As has been stated, there has been no consensus of opinion regarding how the foreign particles act. Illustrations of that great weakness of the theory are given in the next section.

b. Some Suggestions Regarding the Action of the Walls and Particles

It has long been known that freezing can be initiated in a supercooled melt by placing in it either a crystal of itself or a crystal isomorphous therewith. Since a melt containing such a crystal cannot be significantly supercooled, the action of such crystals is not of present interest.

However, it has long been known that other kinds of foreign particles and that "critical" spots on the walls of the container may initiate freezing. It is the various ways in which such particles and spots have been supposed to act that are of prime interest in the present discussion. They fall into three distinct groups: (1) the action is a purely mechanical one; (2) the particle or spot acts solely as a base upon which the molecules of the melt are laid down in crystalline array; and (3) the occurrence of freezing depends upon the action of the adsorbed layer covering the particle or spot. Several kinds of action have been suggested. In this connection, an article by Lehmann [50] (1900) is of interest.

The following are the more important proposals that have been made regarding the manner in which particles and walls initiate freezing:

(1) *Mechanical.*—The earliest suggestion [22, 23] seems to have been that a foreign particle and the wall of the container severally exert a retarding action upon the thermal motions of the molecules of the melt that are adjacent thereto. Thus arises a region in which the molecular velocities are lower than they are at more distant points; i.e. the effective temperature in such a region is lower than that elsewhere, and consequently, freezing, in accordance with the homogeneous theory, will occur more readily there. Ostwald [62], however, seemed to have thought that there might be some other type, or types, of action.

(2) *Direct Deposition.*—Not a few seem to have regarded the particle as serving merely as a substratum on which the molecules of the melt are laid down, one by one, and are held by the attraction of the particle, that attraction being strong enough to retain the molecules, but not so strong as to prevent them from taking up the orientations and relative positions demanded by the structure of a crystal of the melt.

(3) *Adsorbed Layer.*—Several [16, 52, 57] have suggested that the adsorbed layer of the melt on the particle serves in some unspecified way to initiate freezing. Others have been more specific. The several types of action proposed by them are individually listed in the following paragraphs.

Interfacial layer incomplete. In 1904 Füchtbauer [36] suggested that the efficacy of a particle in the initiation of freezing depends upon the extent to which the condition of the adsorbed layer departs from that corresponding to equilibrium. Thirty years later Fricke [35] offered a similar suggestion. Neither indicated how such departure enabled the particle to become effective. The suggestion seems to have been made for the purpose of explaining the facilitation of supercooling by either heating or aging the specimen, as such treatments would probably facilitate the attainment of equilibrium. It was suggested that the particles are colloidal.

Catalytic action. In 1922 Hinshelwood and Hartley [44] concluded that the effective particles are colloidal organic dust from the air, and that the initiation of freezing is a heterogeneous catalytic process, the efficacy of the particle depending in part on its radius. No details were given.

Crystalline adsorbate. In 1932 William T. Richards [67] suggested that a particle is effective in virtue of its being a carrier of the melt in the form of a "crystalline adsorbate," which he seems to have regarded as consisting of crystals of the melt. But four years later [68] he used the same term to denote merely an adsorbate of such a kind as will initiate freezing when brought into contact with the suitably supercooled melt. The nature of that adsorbate is not more definitely stated, nor is the way in which it functions particularized. The preceding refers to those particles whose efficacies are reduced by a prolonged heating of the melt.

He thought that those particles whose efficacies are not affected by heating functioned in some other, but unspecified, way.

(4) *Crystal Lattice.*—In 1938 Hammer [43] suggested that the particles are effective in virtue of their being more or less covered with a thin crystal lattice of the melt. Presumably, the action was regarded as similar to that of a crystal of the melt, differing therefrom only by the effect of the forces of adsorption.

(5) *Spots on Walls.*—That freezing may be initiated at certain fixed "critical" spots on the walls of the vessel containing the melt has long been known [see 74]. How do these spots differ from the rest of the wall surface? What peculiar conditions, if any, other than those already considered for particles, enable them to initiate freezing?

It was early observed that the surface was frequently scratched or rough at those spots. In such cases, the attendant slight elevations of the surface have quite generally been regarded as equivalent to particles stuck to the wall, and have been supposed to act in the same manner as freely suspended particles. In other cases it has frequently been assumed that the same conditions exist but that the scratches or roughnesses are too slight to be seen. Meyer and Pfaff [58] suggested that dissolved impurities might become precipitated at porous and rough spots on the walls, and that the precipitated material might in some way serve as an initiator of crystallization. Richards, Kirkpatrick, and Hutz [68] suggested that a minute wedge-shaped crevice might serve as a critical spot, its activity arising from a suitable orientation of the molecules of the melt adsorbed on the walls of such a crevice.

c. Phenomena Demanded by this Theory

Some of the phenomena that are demanded by the heterogeneous theory are listed below, and in a later section will be compared with pertinent experimental data. It is obvious that all nominally identical specimens that are each so large as to contain a fair sample of the substance and its heterogeneities will behave alike and in the same manner from time to time, provided that neither the substance nor its heterogeneities undergo any change; but in practice it is found that an actual specimen is seldom or never so large, and that secular changes frequently occur. Consequently, in the following it is assumed (1) that the specimen does not contain such a fair sample, and (2) that during the period under consideration neither the substance nor its heterogeneities undergo any change that is not explicitly mentioned.

1. A specimen, when cooled so slowly that it is always essentially in thermal equilibrium, freezes spontaneously, time after time, at a certain fixed temperature (t_{sf}), which may be called its spontaneous-freezing-point, and is, in general, lower than the melting point of the substance.

2. When a large number of nominally identical specimens are so cooled, the several specimens will in general freeze spontaneously at different temperatures; and those temperatures may be spread over a wide range.

3. When a large number of nominally identical specimens are plunged at the same instant into a cold bath, no specimen will freeze until the temperature of its coldest portion shall have dropped at least to the t_{sf} of the specimen. All those specimens having a t_{sf} at or above the temperature of the bath will have frozen by the time the specimens shall have come to thermal equilibrium with the bath; the others will not freeze at all.

4. The temperatures at which two specimens, differing in volume, but otherwise nominally identical, freeze spontaneously bear no necessary relation to the sizes of the specimen.

5. If each particle gives rise to but a single embryo, which presently envelops it, then at a given temperature a particular specimen can produce only a limited and definite number of viable embryos, one for each particle that can initiate freezing at or above that temperature. As the temperature is lowered, the number increases until it is equal to the total number of potentially effective particles contained in the specimen. Beyond that number it cannot go. Furthermore, when the temperature is kept constant, the number of embryos that become visible within a given time, measured from the instant that the specimen was placed in the bath, will increase with the time until it reaches a definite maximum—the number of particles that can initiate freezing at or above that temperature. The increase will be at first rapid and then ever more slow, becoming zero when the maximum is reached. This variation is due in part to the differing temperatures at which the several particles can initiate freezing, in part to the time required for an embryo to become visible, and in part to the fact that the initiation of an embryo depends upon the occurrence of a favorable impact of a molecule of the melt with the particle, which involves an element of chance.

4. COMPARISON OF OBSERVATIONS WITH THEORIES

Several types of observations are listed below and compared with the demands of each of the two theories of freezing. They have been chosen for the purpose of enabling one to appraise the conflict between them and the theories, and between the two theories.

a. Paucity of Singularities

Several observers have commented upon the vanishingly small number of effective singularities as compared with the number of molecules in the specimen. Tammann [78] has stated that not more than one thousand singularities per cubic millimeter become visible after an exposure of 1 minute to any low temperature.

On the homogeneous theory, such paucity indicates that an exceedingly rare, almost vanishingly rare, type of molecular collision is required for the initiation of a viable embryo if the molecules involved are the ordinary molecules of the liquid. In fact, Schaum and Riffert [73] concluded that it is inconceivable that the necessary number of molecules can by mere chance come together in the proper manner. Various suggestions for avoiding that difficulty have been proposed. Some call for special molecules [71, 80, 81, 82]—polymers [10, 11, 12], crystal residues [34, 88, 96], anisotropic molecules [63, 81, 83, 85], etc. Some suggest that singularities arise only in abnormal regions in the melt—regions in which the relative velocities of the molecules are abnormally low [85], in which the molecular packing is denser than usual [34], in which the arrangement of the molecules is more orderly than elsewhere [33, 71, 72], etc. Obviously, each of these restrictions, reducing as it does either the number of suitable units, or the actual volume of melt involved, or both, makes less startling the paucity of the singularities. But at the best, these suggestions are after-thoughts, are makeshifts, that mitigate rather than remove the difficulty. Although they may in certain cases accord with the existing conditions, there seems to be no experimental evidence that any óf them do so accord.

On the heterogeneous theory, the paucity of the singularities is not at all surprising. Their number is simply that of the number of foreign particles that are effective at the temperature considered, and that number may well be small.

b. Supercooling

As shown in this report, a given specimen of water may be markedly supercooled, time after time, to a certain fixed limiting temperature at which it always freezes spontaneously. It will not freeze spontaneously above that temperature, and it cannot remain unfrozen below it. Similar observations have been reported by others for other substances [36, 41, 45, 74, 84, 98].

Such behavior conflicts with the demands of the homogeneous theory, but is readily accounted for on the heterogeneous one.

c. Effect of Volume of Melt

If two specimens differing in volume but otherwise nominally identical be plunged simultaneously into a cold bath at a certain fixed temperature, the smaller may, on the average, freeze more quickly than the larger one [16, 23, 30, 36].

That is contrary to the demands of the homogeneous theory, but not at all inconsistent with the heterogeneous one. On the latter theory, such an observation means no more than that the smaller specimen happens to have one or more particles that will initiate freezing at a higher temperature than will any that happen to be in the larger one.

d. Heterogeneity of Melt

As shown in this and the previous report [28], a given sample of water may be quite heterogeneous, two specimens taken from it behaving consistently in different ways. Such heterogeneity has been reported by others for other substances [36, 68].

On the homogeneous theory, a melt cannot be truly heterogeneous. Such spurious heterogeneity as may result from fluctuations in the state of the molecules would be continually changing, with the result that now one specimen will freeze the more readily, and now the other.

On the heterogeneous theory, such heterogeneity as has been observed is to be expected.

e. Nominally Identical Specimens Differ

As shown in this report for water, and as has been reported by others for other substances [36, 43, 44, 68, 70, 95], nominally identical specimens may exhibit wide differences in behavior. Some will freeze spontaneously at high temperatures and some at low. When plunged into a cold bath, some will freeze quickly, others after a longer interval, and some may not freeze at all. Furthermore, in all these cases the several specimens exhibit marked and enduring individual characteristics, the same individuals always falling into the same class.

On the homogeneous theory, the specimens would at any instant be expected to differ, the differences resulting from chance differences in the molecular encounters. But a specimen should *not* show any individually characteristic behavior; if it froze quickly at one time, it should at another time require longer to freeze under the same conditions.

On the heterogeneous theory also, the specimens would be expected to differ as a result of chance differences in the activities of the particles they contain. But each specimen would be expected to exhibit a characteristically individual behavior.

f. Effect of Preheating

Not only here but elsewhere, and for various substances, it has been found that a prolonged preheating of the melt may facilitate its supercooling. The effect is usually permanent; i.e., it is not annulled by a subsequent freezing [16, 23, 24, 25, 36, 44, 63, 70, 73, 74, 86, 92, 95, 96]. However, Young and Burke [95] state that "to a great extent" the effect "is not permanent."

It is obvious that the effect of heating is to destroy, or to render ineffective, some kind of structural entity that would otherwise remain essentially unchanged over long periods at temperatures that are some tens of degrees above the melting point of the substance, those entities being in some way capable of initiating freezing in a sufficiently supercooled melt. And in general, the entities are neither restored nor rendered effective again by any number of subsequent freezings and meltings, nor by holding the substance for long periods at temperatures which are either above or below its melting point.

Four distinct suggestions have been made regarding those entities: (1) They are enduring modifications of the molecules of the melt—"anisotropic molecules"— remaining from a previous melting of the frozen substance [63, 73, 81, 85].

(2) They are enduring structures composed solely of molecules of the melt. Some [39, 40, 71, 96] have called them precrystalline structures; others [17, 39, 72, 92, 96], regarding them as minute undisintegrated aggregates left from a previous melting of the frozen substance, have called them "residues" of crystals.

Prolonged heating at a suitable high temperature is supposed to destroy those structures.

(3) They are foreign particles that carry an adsorbed layer of the melt. Some [43, 67] regard the layer as composed of actual crystals of the melt; others [16, 44, 68] merely assume that the layer, whatever it may be like, is of such a kind that it will initiate freezing in a suitably supercooled melt in contact with it. The ability of those layers to persist at temperatures above the melting point of the substance and their inability to initiate freezing unless the melt is suitably supercooled are each attributed to the restraints imposed by the forces of adsorption. Prolonged heating at a suitable temperature is supposed to destroy those layers.

(4) They are foreign particles carrying an adsorbed layer that is not in equilibrium; the ability of such a layer to initiate freezing in a supercooled melt in contact with it depends upon the extent of the departure of the layer from its state of equilibrium, and increases with that departure. Heating is supposed to facilitate the attainment of equilibrium, and in that way decreases or destroys the power of the layer to induce freezing [36].

Of these four suggestions, the first two alone can in any way be regarded as compatible with the homogeneous theory of freezing; and that compatibility is distinctly forced, because the presence of such structures, even though composed of molecules of the melt, makes the system a heterogeneous one. The other two are obviously means for adapting the heterogeneous theory to this effect.

It is also obvious that the entities considered in the first three suggestions should be, at least in part, restored by subsequent freezings and meltings. But most, perhaps all, of the available evidence suitable for the purpose indicates that they are not so restored. That of itself should be a sufficient reason for rejecting those suggestions.

But there are other good reasons for rejecting the second. As pointed out in an earlier section, Gibbs has shown that, for any given temperature of the melt, any such structure as is postulated in that suggestion must be of a certain "critical" size if it is to be in equilibrium. Structures larger than that will grow, ultimately becoming crystals; those smaller will decrease continuously to evanescence. The higher the temperature, the larger is the critical size. Consequently, a structure that is of the critical size at a temperature just below the melting point, will decrease to evanescence when the temperature is raised above that point. Furthermore, it is known experimentally that a very small crystal has a lower melting point than does a larger one. That is in accord with Gibbs's conclusion; and one is forced to conclude, in the absence of very strong evidence to the contrary, that still smaller structures and any that can fairly be called a "residue" of a crystal will be quite unable to endure at temperatures above the melting point of the substance. The second suggestion must, therefore, be rejected.

But note! Nothing that has just been said implies that such structures, "residues," or very minute crystals may not have a temporary existence at temperatures above the melting point. Melting is not an instantaneous act. Those structures may so exist for an appreciable time, but throughout that time they are steadily decreasing in size, and they ultimately vanish.

Furthermore, the structures required for explaining this effect must be lasting ones. They should not be confused with any of the numerous transient structures, or pseudostructures, that may arise fortuitously from molecular agitation—structures that are in dissociative equilibrium with the rest of the melt.

Of the structures postulated in the fourth suggestion, little can be said. It seems that no one has explained how a departure of an adsorbed layer from equilibrium can confer upon it the ability to initiate freezing, nor how the extent of that departure can govern the extent of that ability, nor why equilibrium should not be rather quickly established at temperatures somewhat above the melting point. The suggestion seems to be merely an *ad hoc* one that may conceivably be appropriate. As it stands, the suggestion seems to have no experimental basis.

Consequently, it seems fair to conclude that the observed effect of preheating a melt has not been satisfactorily explained on either theory.

g. Effect of Filtration

Numerous observers [36, 46, 52, 56, 58, 62, 96] have reported that a careful filtering of a melt may facilitate its supercooling, the effect being permanent. The same is true of centrifugalizing it [16].

Such an effect cannot be accounted for on the homogeneous theory; it must be dismissed as an irregularity caused by the presence of a few casual heterogeneities. But on the heterogeneous theory the effect is what should be expected.

h. Effect of Duration of Cooling

As shown herein, the length of time that a specimen has been kept at a supercooled temperature has no effect upon the temperature at which it freezes spontaneously. Similiar observations have been reported by others [36, 84].

On the homogeneous theory, the chance of a specimen freezing at a given temperature, below its melting point, increases with the time that it has been kept continuously at that temperature. And if in a large number of trials it be found that the mean time required for the specimen to freeze is τ, then τ is a continuous function of the temperature. Hence, if, when the temperature is lowered by the amount Δt, τ remains essentially unchanged, then when the temperature is increased by the same amount the change in τ should not be large.

Not only is the "chance" required at variance with the observations, but it is very difficult to reconcile the observations with a continuity of τ with t. For example, a specimen of water that froze promptly at $-13°C$ and at every lower temperature, remained continuously unfrozen for over three hundred days while the temperature lay continuously between -8 and $-10°C$ (see table 6). And, when a substance of low viscosity remains continuously supercooled for years without crystallizing, I find it hard to believe that its permanence is to be accounted for solely by an excessively small probability that the molecules will arrange themselves in the crystalline structure. Until forced to the contrary, it seems to me to accord far better with sound practice, as well as to be more reasonable, to assume that, barring a change in the specimen, the probability of its crystallizing at that temperature is actually zero, not merely exceedingly small. Illustrations of such long supercooling may be found in de Coppet's paper of 1907. He reported that a specimen of salol (m.p. $42.5°C$) had been kept supercooled (apparently at 7 to $12°C$) for six years at the time of writing, and that solutions of $Na_2SO_4 \cdot 10H_2O$, saturated at $31°C$, had been kept at 5 to $13°C$ for over twenty-five years, and another lot for over thirty-three years, without any crystallization [23].

On the other hand, the observations do not conflict with the heterogeneous theory. That theory gives no reason for supposing that the duration of cooling would have any significant effect upon the freezing of a melt containing unchanging heterogeneities. There might be a minor, secondary effect depending upon the time that necessarily elapses between the instant at which freezing is initiated, and that at which the crystals become clearly visible, but for a substance having such a high linear velocity of freezing as has water, that elapsed time would probably be very short. However, if the temperature were lowered so very abruptly that freezing occurred before the adsorbed layer had become adjusted to the temperature of the melt, freezing might possibly occur at a higher temperature than would otherwise be the case.

i. Rate of Nucleation

Tammann [78], accepting the homogeneous theory, which demands that the number of singularities of a specified type that arise per unit time in a given specimen of melt shall be a constant, characteristic of the substance and its temperature, called the ratio of that number to the volume of the melt the "Krystallisationsvermögen," and most frequently the "Kernzahl," of the substance at the temperature considered. As neither of those terms indicates clearly that the quantity considered is a rate of production, it has seemed well to replace them in this report by the expression "rate of nucleation." Tammann and his disciples have at-

tempted to determine that rate and its variation with the temperature of the melt.

Since the singularities themselves are entirely invisible, their number has to be inferred from that of the resulting embryos that grow to a visible size. That is, the types of singularity that are involved are those that give rise to embryos that are viable under the existing conditions. In order to avoid certain serious experimental difficulties, Tammann adopted a procedure involving a quick transfer of the specimen from the bath in which the singularities were formed to another at a higher temperature, at which no additional singularities arose, and the embryos formed at the lower temperature then grew to visible size, and were counted.

As pointed out by Volmer and Weber [89] in 1926, and by others later [14, 16, 59], this procedure introduces an unavoidable error. At the instant that the specimen is transferred from the lower to the higher temperature, it contains embryos of various ages and sizes, some that have just been formed being no larger than the critical size corresponding to the lower temperature. When the specimen containing this mixture of embryos, all viable at the lower temperature, is raised to the higher temperature, only those embryos that are at least as large as the critical size corresponding to that higher temperature will continue to grow. The others will continuously decrease to evanescence. Consequently, the number counted will be less than the number of viable embryos formed at the lower temperatures. Furthermore, the higher temperature remaining the same, that defect will be the greater, the lower the temperature at which the singularities are formed. Hence the variation of the rate with the temperature is not represented by the experimental curve so obtained. Consequently, neither the nominal rate of nucleation nor the nominal variation of that rate with the temperature, as determined by Tammann's procedure, can be relied upon in testing the validity of either theory of freezing. They will not be further considered.

But it would seem that, for any given pair of temperatures, the number of embryos that fail to become visible at the higher temperature will be independent of the time τ during which the singularities were being formed at the lower one. That is, if the number of embryos counted after an exposure of τ be plotted as an ordinate against τ as abscissa for each of a series of values of τ, then the resulting curve will differ from the correct one only by a bodily shift parallel to the axis of ordinates. Hence the form of that curve is a true characteristic of the substance, and can be used with confidence in testing the theories.

It seems that Othmer [63] (1915), working in Tammann's laboratory, was the first to describe that curve. The next to study it seems to have been Biilmann and Klitt [16] who published such curves in 1932. Hammer [43] published similar curves in 1938. All these curves are of the same type. At any given temperature the number of viable embryos produced within an interval τ increases monotonously with τ, at an ever decreasing rate, to a fixed maximum, characteristic of the specimen and the temperature. The lower the temperature the higher is that maximum. The time τ required for attaining the maximum is short and varies but little with the temperature.

Those curves are entirely incompatible with the homogeneous theory in at least two particulars. The theory demands that the number should vary linearly with τ, increasing without limit, whereas the observations show that the number does not vary linearly with τ, and that it soon reaches a fixed limit.

Those curves are not inconsistent with the heterogeneous theory. But they do demand that the theory shall require (1) that, at any given temperature, no singularity shall give rise to more than a fixed number of viable embryos; and (2) that the singularities shall in general so differ among themselves that they may be assorted into classes such that none of those in a given class can give rise to a viable embryo unless the temperature lies at or below a certain value, differing from class to class, and that those in any class shall, at a given temperature, differ among themselves in the facility with which they give rise to viable embryos. It would seem that such demands can probably be met by one or more of the proposals that have been made regarding the way the particles act in initiating freezing.

j. Mechanical Initiation of Freezing

It has long been known that the freezing of a supercooled melt may, at least in certain cases, be induced by mechanical means.

Neither of the theories being considered offers a satisfactory explanation of this phenomenon.

k. Preferred Temperatures

Neither of these theories suggests, or offers any explanation of, the existence of such preferred temperatures as have been revealed by the observations herein reported.

5. CONCLUSIONS

Of these eleven observed phenomena, six (b, c, d, e, g, and h) are entirely incompatible with the demands of the homogeneous theory, but present no difficulty to the heterogeneous one; three (f, j, and k) have not been satisfactorily accounted for on either theory. Of the remaining two, a presents no difficulty to the heterogeneous theory, but its explanation in terms of the homogeneous theory is at best merely a plausible one; and c is incompatible with the homogeneous theory, but probably can be explained satisfactorily on the heterogeneous one. More briefly, the homogeneous theory is directly antagonistic to seven of these phenomena, offers no explanation of three, and its proposed adjustment to the remaining one is merely plausible; the heterogeneous theory can readily account for seven,

offers no explanation of three, and can probably account for the remaining one.

Whence it seems that a new theory of freezing is needed, one that is much closer akin to the heterogeneous than to the homogeneous theory, but that does not completely exclude the latter. It must account, without special assumptions, for all eleven of the phenomena just considered, and, so far as it applies, for all those other observations reported earlier in this paper on the freezing of water, and it must offer a rational explanation of how foreign particles may give rise to embryos, without ascribing to those particles any property especially invoked for that purpose. Furthermore, it should be applicable, with only minor and obvious changes in the wording, to the crystallization of solutes from their solutions.

Such a new theory will be outlined in a later section, but before turning to that, it is necessary to consider seriously how it happens that the homogeneous theory fails so signally to account for the observed phenomena.

IV. FAILURE OF THE HOMOGENEOUS THEORY

How does it happen that the homogeneous theory of freezing, as heretofore understood, based as it appears to be on the kinetic theory of matter, fails so signally to account for the observed phenomena? Must it not arise from discrepancies between the actual conditions and the assumptions on which the theory rests?

There seems to be no reason for doubting that, as stated in the preceding discussion of the theory, freezing will begin "at those places where such suitable fortuitous encounters of molecules of the melt occur that the colliding molecules can become so bonded together as to form a persistent unit" which can grow and presently become a visible crystal. However, the questions arise: "Can such suitable encounters occur under the actual conditions?" and "Are the encounters fortuitous?" The theory tacitly assumes that the answer to each of these questions is "Yes." Is that assumption justified?

A fortuitous event is one that happens by chance. This does not mean that the event could not have been foreseen by one acquainted with all the pertinent situations leading up to it, but only that the effect of those situations is in no way causally related to any of the corresponding ones that determine other similar events in the system in which one is interested. That is, each event in that system is entirely independent of every other event of the same kind; as in the case of successive throws of a true die. It is such independence that justifies the use of the words "chance" and "fortuitous."

Such independence does not exist between the successive states of the molecules of such a system as that with which one is here concerned; viz., a finite volume of melt in a closed vessel, the whole being in thermal equilibrium at an unchanging finite temperature (as those terms are commonly understood), and unacted upon from without. Such a system is a closed one, and is .

absolutely determinate. The state of the molecules in any given small volume of it at a given instant is fully determined by the state of the system at that preceding instant which is chosen as the fiducial one. Although neither you nor I can specify the state of the molecules in that volume at the specified instant, that state could be fully specified by one who knew in detail the fiducial state of the system, and who was able to comprehend the complete system in all its immense complexity. In the system with which we are concerned, the nature of the encounter of molecules at a given point and time is not a fortuitous event. The first tacit assumption heretofore mentioned is invalid.[8]

Only certain situations can arise; and it may well be that no molecular encounter of the kind required for the initiation of freezing can occur, unless it be at very low temperatures.

The commonly held contrary belief seems to rest on the second tacit assumption heretofore mentioned; viz., that, in the system with which we are concerned, each one of the factors—speed, orientation, density of packing, etc.—involved in specifying the nature of the encounter of molecules at a given point may, and in the course of time will, take every conceivable real value.

That such is not the case for the molecular speed is readily apparent. Since the absolute temperature of our finite system is by hypothesis finite, and the total kinetic energy of translation of its molecules ($\frac{1}{2}\Sigma mv^2$) is likewise finite, no v can be truly infinite. Here m is the mass of a single molecule, v is that molecule's linear velocity of thermal agitation; and Σ denotes summation over all the molecules of the system.

Should one be inclined to object to this conclusion on the ground that an infinite speed is demanded by an expression derived in the mathematical study of the dynamical theory of gases—which serves as the foundation of the homogeneous theory of freezing—he should recall that no mathematical procedure can reveal anything that is not involved in the fundamental postulates —all unproven—on which the procedure rests. If without error the procedure leads to an expression demanding an infinite speed, then one or more of the postulates were of such a nature as to require it; and if an infinite speed is impossible in the physical system under consideration, then the *ensemble* of those postulates is invalid for that system, and at least some of the conclusions that may be derived from them will likewise be invalid. Obviously, the difference between a situation being infinitesimally probable and being nonexistent is truly infinite. One cannot validly be equated to the other, not even in those cases in which it is experimentally impossible to tell which is which.

---

[8] Buerger [20] has written: "Since energy is communicated to an atom mechanically by its neighbors, it can be said that the motions of the atom are not entirely at random, but are synchronized, in a general way, with those of their neighbors." He was writing of crystals, but the statement would seem to apply also, in a general way, to the molecules of liquids.

Since the several mathematical treatments of the dynamical theory of gases all lead to an expression demanding that the molecular speeds shall cover the entire range of real numbers, whereas no molecule in a finite mass of gas at a finite temperature can have an infinite speed, it is evident that no conclusion arrived at by such treatments should be regarded as more than approximately applicable to such a mass of gas.

Indeed, it seems that in every practical application of those conclusions to gases the tails of the distribution curve—those portions corresponding to exceedingly small probabilities—are always ignored. But it is exactly to those tails—for which there seems to be no experimental evidence whatever—that the adherents of the homogeneous theory of freezing appeal in their attempts to explain the paucity of singularities and the existence of supercooling, and to justify their opinions that the spontaneous-freezing-point of a given specimen can have no fixed value, that a large volume of melt will freeze more readily than a smaller one, and that the length of time that a specimen has been kept supercooled to a given temperature is an important factor in determining whether it will freeze.

In fact, I know of no experimental evidence that any of the various elements involved in the specification of the state of the system varies over all conceivable real values, including those which have commonly been regarded as infinitesimally probable, but not as nonexistent. It may well be that no such molecular encounter as is required for the initiation of a viable embryo can ever occur except perhaps at and below some very low temperature. Only by experiment can one determine what that limiting temperature for any given substance is. The observations herein reported indicate that for water it lies below the lowest temperature ($-22°C$) investigated.

In this connection it is interesting to recall that as early as 1878 J. W. Gibbs [38: 255–257] pointed out that the initiation, in the interior of a preexistent homogeneous phase, of a new phase of such a kind that the initial fault in the homogeneity is "small in extent but great in degree" is impossible unless the temperature and pressure are such that the equilibrium size, as defined by him, of the globule containing the fault is zero. And that "the real stability of a phase extends in general beyond that limit . . . at which the phase can exist in contact with another at a plane surface." That is, in general, a supercooled homogeneous melt is truly stable at least down to some temperature which cannot be exactly specified by his procedure, but which may be very low.

It should, however, be remarked that nothing in the preceding discussion denies that it is theoretically possible to assemble molecules to form a finite volume of liquid having any desired distribution of molecular speeds, etc., even including an infinite speed or a distribution that would lead to the initiation of freezing. However, the first is incompatible with the specimen's being at a uniform and finite temperature; and the second, except possibly at very low temperatures, is incompatible with the requirement that the state of the specimen is such as can be attained by a mere cooling of the specimen slowly from a condition of equilibrium at a higher temperature. In general, such arbitrarily constructed specimens would not be in thermal equilibrium.

The failure of the homogeneous theory of freezing, as heretofore understood, to account for many observed phenomena arises from the invalidity of the assumptions on which that theory rests. The past unquestioning acceptance of those assumptions seems to have resulted from an unwarranted reverence for physical conclusions announced by mathematicians, and from a confusion of ideas, from a confusion of "what I cannot foresee" with "what cannot be foreseen" and of "all conceivable situations" with "all situations compatible with the actual conditions of the problem." In each case the confusion led to an ignoring of constraints inherent in the problem.

V. A NEW THEORY

1. INTRODUCTION

Recalling that the observations reported herein are inconsistent with the assumption that the observed spontaneous freezing of water in the range 0 to − 20°C is initiated at homogeneous singularities; that the formation *per saltum* of a large embryo by means of a multiple collision of free molecules involves difficulties, perhaps insuperable; and that Gibbs [38: 255–257] has shown that, if isolated singularities can form in the interior of a homogeneous phase, it is in general only at temperatures well below the melting point of the substance, a homogeneous melt being in general stable well below that point; recalling these things, one seems justified in assuming that it is probable that the homogeneous theory is applicable only at low temperatures, if at all.[4] That will be assumed.

Hence the new theory, if based on the kinetic theory of matter, as it is, must be a heterogeneous one. But it should not demand that the singularities shall be of any special kind, although it will recognize that very special kinds may exist, and will produce corresponding effects. It should be applicable at all temperatures and to all melts, and with obvious minor adjustments, it should equally apply to the crystallization of solutes from their solutions. It must provide satisfactorily for the occurrence now and again of groups of enough

molecules having small relative velocities to suffice for the formation of embryos of viable size.

It should be remarked that the term "catalytic," to the extent that it involves the idea of a speeding up of a process that would otherwise proceed at a slower rate, is not properly applicable to the action of these singularities in the initiation of freezing. They do not speed up the freezing; they merely initiate it. In their absence the melt would not freeze however long one waited.

The one thing that is common to every heterogeneous singularity is the adsorbed layer always found at interfaces between differing substances. There, molecules of the melt are assembled, and from that layer it should be possible, by molecular impact and other means, and under suitable conditions, to loosen such groups of molecules as are needed for the formation of viable embryos.

That is the basic idea of the New Theory. The theory assigns to the adsorbed layer, merely as an adsorbed layer and irrespective of any other special characteristic, the supremely important role of assembling molecules and providing such groups of them as are needed for the formation of viable embryos.

In the following pages the theory is developed in outline for ordinary adsorbed layers. Its extension to layers having special characteristics should not be difficult, but each special characteristic will, in general, require a special treatment.

2. THE THEORY

a. Theory Stated

It is generally agreed that the molecules of a liquid that are immediately adjacent to a foreign body are packed quite closely together, are bound quite strongly to the body, and have a preferred orientation with reference to the interface between the body and the liquid. In successively more distant layers, the packing becomes less close, the binding weaker, and the orientation less complete, until presently the status characteristic of the liquid in bulk and far from foreign bodies is attained.

For simplicity, this entire group of layers, whether many or few, in which the arrangement of the molecules is modified by the presence of the foreign body is, in this report, called the *adsorbed layer* of liquid.

If the adsorbed layer can be so torn that a sufficient number of loosened molecules remain in one another's field of force for such a time and with such a degree of freedom that they can reorient themselves and become bound together in the manner characteristic of an embryo that can persist and grow at the existing temperature, they will do so.

This new theory assumes (1) that the adsorbed layer can be so torn by various means, including the impact of free molecules of the melt, and (2) that both the spontaneous and the mechanically initiated freezing of a

---

[4] Rau has reported that the freezing of a small drop of water began repeatedly at a single point and at a reproducible temperature. Presently that point ceased to be effective. As the temperature was lowered further, presently another point became active; and so on. But in no case could he cool a droplet below − 72°C, at which temperature freezing began at numerous points throughout the drop. That ice melted at − 70°C [65a]. *See also* F. C. Frank [31a].

supercooled melt may be, and at least at the higher temperatures are, initiated solely in that manner.

If the adsorbed layer can be so damaged by molecular impacts as to give rise to viable embryos, then the spontaneous freezing of a supercooled melt can obviously be initiated in that manner. And if the layer can be so damaged by molecular impact, it can surely be so damaged by other means, including gross mechanical ones.[5]

Consequently, if satisfactory reasons can be given for supposing that molecular impacts can produce the assumed effects, and if an investigation of the variations in the size of the resulting embryo with the curvature of the interface and with the temperature shows that they are of a kind to account for all the observed phenomena to which the theory is pertinent, then the proposal of this theory is justified.

b. Effect of Speed of Impinging Molecule

It is well known that, other things being the same, the firmness with which an adsorbed layer of a given liquid is held to the foreign substance forming a boundary depends upon the nature of that substance. For some substances the binding is very strong, for others exceedingly weak. In all cases, the outer layers of molecules—those distant from the foreign substance—are very weakly held. Whence it is not unreasonable to expect that there are adsorbed layers that can be damaged by molecular impacts. That will suffice for the present.

If such a layer be suitably struck at a given point by a molecule of the liquid, there will be little or no damage if the speed of the molecule is small; at a greater speed, some of the adsorbed layer is so loosened that an embryo can be formed; as the speed is increased, more and more of the layer is loosened, permitting the formation of larger and larger embryos; but at the same time the tendency of the loosened molecules to scatter is increased, so that presently the number of molecules that are loosened and left with small relative velocities decreases, and, consequently, so does the size of the resulting embryo; and as the speed of the impinging molecule is still further increased, a value is presently reached at which the dislodged molecules are so widely scattered that no embryo can be formed from them.

That is, under any given conditions, including the line of flight of the impinging molecule, the speed of that molecule must lie within a certain limited range if the impact is to give rise to an embryo; and there is a unique speed that results in the production of the largest embryo that can possibly be produced by molecular impact under those conditions.

Furthermore, as the direction of the line of flight of the impinging molecule is changed, all else remaining the same, both the unique speed and the size of the corresponding embryo will, in general, change. Call the very largest of all these largest embryos the *characteristic embryo*, and the corresponding unique speed, the *characteristic speed*, for that portion of the adsorbed layer. For each point of an interface between a melt and a foreign substance there is a characteristic speed, and a characteristic embryo which is larger than any other that can arise from a molecular bombardment of that point. Obviously, the point's characteristic speed and embryo are each independent of the actual temperature of the melt; except as that may affect the binding of the adsorbed layer.

c. Effect of Convexity of Interface

Since molecules can escape more readily from small drops than from larger ones, it is valid to infer that the ease with which a given adsorbed layer can be damaged will increase with its convexity. Furthermore, it is obvious that an impinging molecule can knock molecules from a highly convex surface more readily than from a flat one. (Consider extreme cases: a molecule resting on a point, and one in a flat layer of closely packed molecules.) Thus one is led to consider the question: How do the characteristic speed and the size of the characteristic embryo vary with the convexity of a given interface?

In that each point of an interface may, for present purposes, be regarded as a segment of a sphere, attention may, for simplicity, be restricted to spherical particles; and it will be further restricted to particles of a single substance suspended in a single pure [6] melt.

In order to visualize the problem, consider a rigidly fixed sphere covered with a closely packed layer of marbles, all of the same size, stuck to the sphere, tightly if the sphere is very large, and ever more weakly as its size decreases; and consider the knocking loose of some of those marbles by the impact of another marble, of the same size, thrown at them.

It seems obvious that for small spheres both the characteristic speed and the size of the characteristic embryo will increase with the size of the sphere; and that for an infinitely large sphere, the characteristic speed will be very great, but the characteristic embryo will surely be small. That is, as the size of the sphere decreases from a very large value, the characteristic speed decreases monotonously, but the size of the characteristic embryo passes through a maximum.

The size of the sphere that gives rise to that maximum embryo may be called the *transition size* of a spherical mote of that substance in that liquid. And that maximum embryo may be called the embryo *proper* to the substances forming the interface from which it

---

[5] It is interesting to note that Frankenheim's [32] explanation in 1860 of the formation of crystals along the line where a glass rod had been drawn along the glass-liquid surface was that the rubbing had removed a layer of something from the glass. He thought that the layer interfered with the wetting of the surface.

[6] Actually, the melt will contain in solution some of the substance of the particles.

arose. Likewise, the temperature at which the proper embryo is in equilibrium with the melt may be called the *proper temperature* (t_p) for those substances.

The size of a proper embryo will, in general, vary with the nature of the substances forming the interface, but will be independent of the temperature, except as that affects the tightness with which the adsorbed layer is bound to the sphere. In what follows, it will be assumed that the size is independent of the temperature.

Since the temperature at which an embryo is in equilibrium with its melt depends solely on the size of the embryo, increasing with that, the relation between the size of a spherical mote and the temperature at which its characteristic embryo is in equilibrium with the melt is somewhat as represented diagrammatically in figure 33. The abscissa (c) of P is the transition size

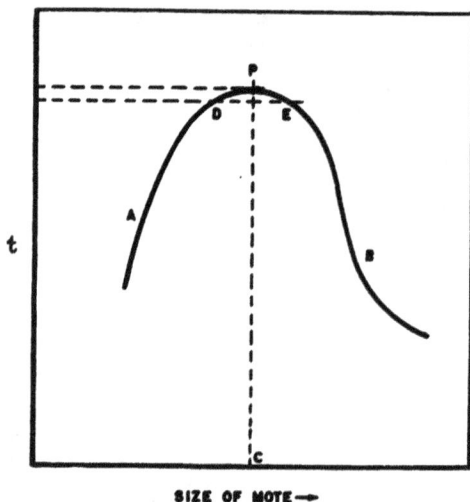

SIZE OF MOTE →

FIG. 33. Size of mote and equilibrium temperature of its characteristic embryo. Schematic only. The mote can yield no embryo that is in equilibrium with the melt at temperatures above t_p. Abscissa C of P is transition size of motes of this substance. Motes smaller than C are minor motes; those larger are major motes.

of motes of that substance in the melt considered, of motes that give rise to the proper embryo; and the ordinate (t_p) of P is the proper temperature. Since the proper embryo is the largest embryo of that melt that can be derived from that mote-substance, it is obvious that under no circumstances can motes of that substance initiate freezing in that melt at any temperature above the proper temperature t_p.

All of which is in agreement with the conclusions reached early in the experimental work (*see* section B, I, 1, o). In conformity with the terminology there adopted, motes larger than c will be called *major* motes; those smaller, *minor* motes.

As a mote decreases in size, the temperature at which it can initiate freezing in a given melt rises if the mote is a major one, and falls if it is a minor one. The mote that is nearest to c in size is always the controlling mote; the melt cannot be cooled below the temperature at which that mote can initiate freezing.

d. Motes and Roughnesses

So far, no limit has been placed on the convexity of the interface, the implication being that embryos may arise from it by molecular impact whatever the convexity may be. That implication is not necessarily valid. It may well be that no embryo can so arise if the interface is either exceedingly convex or very flat. In each case, such failure arises from the impossibility of detaching by molecular impact enough molecules with sufficiently small relative velocities to form an aggregate that can maintain its identity. In the case of very flat surfaces it may be that even at the melting point of the substance there is in the melt no molecule with sufficient velocity to disrupt the adsorbed layer.

But how flat is very flat? Being concerned solely with molecular magnitudes, one may conclude with much confidence that a surface having a radius of curvature 10,000 times as great as the radius of the adsorbed molecule may for all present purposes be regarded as a plane.[7] Since the radius of a molecule of water is of the order of 10^{-8} cm, 10,000 molecular radii is, in that case, of the order of 1 micron (1 micron $= 1\mu = 0.001$ mm). Whence one may conclude that a sphere of 1μ radius is for present purposes not to be distinguished from a truly flat surface. All convexities with which one is concerned in this study are very great indeed.

Where shall one look for such highly convex surfaces as are required? Primarily, to the very small motes that are always suspended in every liquid unless extreme care has been taken to remove them. And secondarily, to rough spots, scratches, and points on the walls of the container and on the surfaces of solids immersed in the liquid.

Since motes can readily become transferred from the liquid to the wall of the container, and back again, and may at times become firmly attached to the wall, then simulating rough spots, scratches, points, etc., it is convenient, now that the latter must be considered, to extend the connotation of the term "mote" so as to cover, in addition to small freely suspended particles of every kind, fixed particles and all kinds of roughnesses, scratches, and points on the surfaces of the solids bounding the liquid. In what follows, the term mote is used in this extended sense.

Since the motes involved in the initiation of freezing are exceedingly small, it seems probable that only solid motes are effective; those of liquid or of gas would, in general, quickly vanish by solution.

---

[7] This corresponds to spheres only 2 mm in diameter stuck to the surface of a sphere 20 m in diameter.

In any consideration of the part played by a mote in the initiation of freezing, important questions regarding the relation of an embryo to its initiating mote arise. Does the growing embryo attach itself to the mote, become frozen to it? Does it completely envelop the mote, forming with it what has herein been called a complex embryo? Does the embryo form an independent free-swimming unit?

If the last is the case, then a mote will be continuously active, giving rise to embryo after embryo, and the rate of nucleation at a given temperature will be a constant, exactly as Tammann and others assumed that the homogeneous theory demanded. However, as has been seen, Tammann's experimental procedure was unsatisfactory, and the published results obtained by others agree in showing that at any given temperature a specimen produces only a definite, limited number of viable embryos. In such cases, it is probable that the embryo either envelops the mote, forming with it a single complex embryo, or merely "freezes" to the mote, making inactive that portion of the mote's surface. Later embryos may then form from other portions of that surface and become frozen to them; until presently the entire surface of the mote shall have become inactive. As embryos can be observed only after they have become visible, it would seem that there is little practical difference between these two cases. The peculiar behavior of the complex embryo should always be kept in mind.

In this connection certain observations made by Frankenheim [32] in 1860 are of much interest. He reported that a liquid seldom crystallizes except in contact with a solid, and that an adjacent free surface of a liquid as well as of a solid, even when not crystalline, has an effect upon the orientation of the crystals. In the latter case the deposited crystals usually have the same face in contact with the solid irrespective of the nature of the solid, being, for example, the same for glass as for feldspar and for metals, and this irrespective of whether the surface of the solid is horizontal or vertical.

No attempt has been made to check these observations, and I do not know whether they have been confirmed by others. But they seem to suggest that the embryo of a given substance does become attached to the adsorbed layer, and in a certain manner. That is, that the embryo usually, perhaps always, forms with the mote a complex embryo.

Whenever embryos adhere to their motes, it should be possible to remove the motes more easily by filtration after they shall have become enlarged by the growing embryo. Meyer and Pfaff [56, 57] have reported such cases; and it is possible that the observations which Barnes [12, 13] interpreted as indicating that the trihydrol of water is removed along with the ice formed in the water are cases of the same kind.

Motes may not only dissolve and adhere to the walls, but they may grow by precipitation of solutes from the melt, they may agglomerate, and those clumps may again break up, and the larger motes may grow at the expense of the smaller. Every such change potentially involves a change in the effective size of the mote. The change may be in either direction, and so may be the change in the spontaneous-freezing-point of the melt. Great and erratic changes in the t_{sf} of a specimen might well be expected, but in fact such changes were seldom observed in the present study of water, perhaps because most of the specimens of water used in this study probably contained a very large number of motes. That would result in an automatic statistical smoothing of the data—a decided advantage in such exploratory work.

If a melt contained only a single spherical mote, it is probable that crystallization starting there would merely grow out into the melt, unless some of the delicate outposts of that growth became broken from the rest; in which case each broken-off portion would probably serve as a "seed" for a new growth of crystals. Observations recently reported by Altberg [2, 4] suggest that such a breaking-off and seeding may frequently occur. However, in the observations herein reported of water containing a bit of cinder, and of aqueous solutions of alcohol and of other substances, the crystallization merely spread progressively from the point at which it started.

I know of no data from which can be obtained a satisfactory answer to any of the following interesting questions: If a melt contained a single and spherical mote, would that mote lead to the inception of a single embryo? If so, would that convert the entire volume of the melt into a single monocrystal? What happens when an advancing crystal meets a mote that is ineffective at the existing temperature? Will it so disrupt the adsorbed layer as to initiate a new embryo? Will it entrap the mote, or reject it? How will the advancing crystal be affected by the encounter? Are Tyndall's "flowers of ice" related in any way to the motes that were contained in the water from which the ice was frozen? If so, how?

e. Temperature of Spontaneous Freezing

It has previously been seen that no embryo is viable unless it is larger than a certain critical size determined by, and increasing with, the temperature of the melt.

Consequently, as the specimen is slowly cooled and so long as any molecule of the melt has the characteristic velocity,[8] freezing cannot occur until the temperature

---

[8] If, before spontaneous freezing occurs, the temperature of the melt shall have been so reduced that none of the molecules has a speed so great as the characteristic speed, then every embryo formed by the impact of a molecule will be smaller than the characteristic one, the largest being that formed by the impact of the swiftest molecule present in the melt. As the temperature is lowered, the size of that largest embryo will decrease; and, if its rate of decrease exceeds that for the critical embryo, spontaneous freezing cannot be initiated by

shall have fallen to that for which the critical size of embryo is equal to that of the previously defined characteristic embryo of the specimen, and the specimen must then freeze within the interval of time required to ensure the required kind of molecular impact upon the proper point or points of the interface. That interval may be, and in the work here reported seems to have been, very short.

Hence a changeless specimen should, on this theory, freeze spontaneously at a definite fixed temperature, and should possess complete thermal stability throughout the range of supercooling that lies above that temperature. That has been found to be true for water. The theory is to that extent entirely satisfactory.

f. Effect of Thermal Shock

If the melt be cooled very rapidly, it may be that the state of the adsorbed layer cannot keep in equilibrium with it. In that case, the actual layer corresponds to a temperature that is higher than that of the melt, and consequently can be damaged more readily than can one that is in equilibrium with the melt. Hence, if the mote is a major one, the characteristic embryo will be larger than if the layer were in equilibrium, and freezing will occur spontaneously at a higher temperature; if the mote is a minor one, the reverse will be true.

Results of the former type have been reported. See e.g. Tammann and Büchner [84].

g. Secular Variation

(1) *Simple Drift.*—If a melt contains only minor motes, its temperature of spontaneous freezing (t_{sf}) decreases monotonously as those motes become progressively smaller by dissolving in the melt, the controlling mote being always the largest one.

The drifts shown by the graphs in figure 3 are qualitatively of the kind that should be expected under those conditions, but only a mote that dissolves exceedingly slowly could give rise to a drift so slow and so lasting.

Such slow solution might be a property of the material of the mote. However, the observed t_{sf} for $C14$ (fig. 7) changed rapidly from one value to another, indicating that, as measured by changes in t_{sf}, the rate of solution of each of the several motes in that specimen was rapid. Perhaps $C14$ contained no mote of the same material as those responsible for the data given in figure 3.

Slow solution might however result, even if the material were readily soluble, provided that the melt was

---

molecular impact at any temperature. If, however, the size of the embryo formed by the impact decreases less rapidly than does that of the critical embryo, then there may be a low temperature at which the two will have the same size; freezing will then occur spontaneously at that temperature. Whatever the situation, one or other of two cases will result: either spontaneous freezing cannot be initiated in the manner considered, or it will occur at a definite temperature.

nearly saturated with it, and it might continue for a very long time if the melt contained a large number of motes of widely varying sizes. The small ones, being more soluble than the large, would be continually dissolving, and the large ones would be as continually growing. That would result in a certain average concentration of the material in the melt, which concentration would slowly change as the smaller motes decreased in size and vanished by solution. If the concentration happened to be such that the melt was saturated with respect to the controlling mote, t_{sf} would be essentially constant for a time; if it were less, that mote would slowly dissolve, and if it were a minor one, t_{sf} would slowly drift downward; similarly for other cases. Situations of that kind may arise.

It should be recalled that $C14$ may be expected to contain relatively few motes and very little dissolved material; whereas of the specimens of figure 3, $C27$ (stagnant pool) and $C33$, $C34$, and $C48$ (Washington City water) may be expected to contain many motes and not a little dissolved material, and the same may be true for $C8$ and $C9$ (ordinary distilled water) and for $P30$ and $P33$ (prepared in the dusty room). Although not necessarily so, it may be that a combination of many motes and a relatively high concentration of mote-material in the melt is the cause of the sluggishness exhibited by the graphs in figure 3—is the difference in the mechanism involved in figure 3 and in $C14$ and remarked upon in an earlier section.

(2) *Preferred Temperature.*—If a melt contains only major motes, t_{sf} is controlled by the smallest mote. As the motes dissolve, decreasing monotonously in size, t_{sf} rises until the smallest mote shall have been reduced to the transition size C (fig. 33), when $t_{sf} = t_p$. Then that mote becomes a minor one, and t_{sf} decreases until such time as it shall have become the same as that for the next smallest, and still major mote. Then that mote becomes the controlling one, and t_{sf} rises to t_p; and the process is repeated. The value of t_{sf} rises to t_p time after time, and between each adjacent two of those values it falls progressively to a lower one, from each of which it progressively rises again to t_p. The particular lower value to which t_{sf} falls depends upon the difference in the sizes of the two motes concerned, and may be expected to vary widely.

If, however, no mote of a group differs in size from the next smallest one by more than a small amount, then t_{sf} will continue to lie near, or essentially at, the proper temperature t_p until the entire group shall have passed the transition size (C) for such motes. Thereafter and as previously described, t_{sf} will undergo a simple drift to lower values, until other motes take over control. In this case, the proper temperature (t_p) is evidently a preferred temperature. As examples of this type of variation, see graphs for $C23$ and $C40$ (fig. 4).

Obviously, the motes that take over control while t_{sf} is drifting may be of the same kind as the preceding and form a similar group. Then t_{sf} will progressivly rise to the same value (t_p) as before, and will remain there until that group shall have passed C.

A specimen may freeze at the same preferred temperature time after time; the successive recurrences may be separated by intervals of any length; and each recurrence may persist for any length of time, long or short, even for so short a time as that defined by the passage of a single mote.

If, however, the group contains relatively few motes and each differs in size from the next smallest one by a truly significant amount, varying irregularly from one pair to the next, then t_{sf} will drift up to t_p, down to some lower value t', up to t_p, down to t'', up to t_p, and so on, the temperatures t', t'', . . . differing irregularly, one from the other, each, however, being by hypothesis lower than t_p. If the time-rate at which t_{sf} varied were the same for every part of the curve of figure 33, one could scarcely draw a definite conclusion regarding the distribution among the several temperatures of the values of t_{sf} that are observed at irregularly distributed instants of time. However, it may safely be assumed that the time-rate at which t_{sf} changes as a mote passes C (at P, fig. 33) is much less than that corresponding to a point on either flank of the curve. Hence, here also the observed values will tend to cluster just under t_p rather than at any other temperature. Even in this case, t_p is a preferred temperature.

If a specimen of water contains motes of two substances (a and b) having the proper temperatures t_{pa} and t_{pb}, t_{pa} lying much above t_{pb}, and if a is present only in the form of minor motes, but b is present as effective major motes as well as minor motes, then as the specimen ages, its t_{sf} will simply drift until it shall have reached t_{pb}, or a little lower, when the b motes will take control. Thereafter t_{sf} will, in general, be independent of the a motes. That is, the t_{sf} of an aging specimen cannot exhibit a simple drift much beyond the transition temperature of a substance which is present as effective major motes. The same is true when the aging is hastened by heating the specimen. Behavior of this kind has been observed, and has been remarked upon in the experimental portion of this report.

Such preferred temperatures—the proper temperatures of the several mote-materials—should not be confused with any persisting constant value of t_{sf} that occurs when the controlling mote remains for a time unchanged in size as a result of a suitable concentration of the mote-material dissolved in the melt. That constant value, depending on a concentration and not solely on the natures of the substances involved, is a strictly casual one, not to be expected under other circumstances, and when once lost by a partial solution of the mote, is never recovered. Nevertheless, there seems to be at present no way in which one can with certainty distinguish between such casual constancy and an unrepeated preferred temperature.

(3) *Multiple Preferred Temperatures.*—For a given melt, the value of t_p varies, in general, with the material of which the mote is formed. Consequently, each specimen of a melt may be expected to have as many preferred temperatures as there are mote-materials in it. Specimens obtained from the same source and treated in the same way should be expected to have one or more preferred temperatures in common. Those obtained from other sources might have quite different sets of preferred temperatures.

Consider a specimen having three preferred temperatures—t_{pa}, t_{pb}, t_{pc}—decreasing in that order. So long as a swarm of pa motes holds sway, $t_{sf} = t_{pa}$; when that swarm has passed its C, t_{sf} will drift to lower values, and if when it reaches t_{pb} a swarm of pb motes takes control, t_{sf} halts there. If no new swarm of pa motes takes control until well after the pb swarm has passed, t_{sf} will on the passage of that swarm drift to still lower values, and may reach t_{pc}, at which it will stop if a pc swarm takes control. Whenever during its downward drift a pa swarm takes control, t_{sf} at once drifts upward to t_{pa}, to stay there until the swarm has passed. Similarly, if a pb swarm takes control while t_{sf} is drifting from t_{pb} to t_{pc}, or is at or below t_{pc}, t_{sf} at once begins to drift toward t_{pb}, and it may be that a pa swarm will take control before it reaches t_{pb}, in which case it will not halt there, but will proceed to t_{pa}. Likewise when t_{sf} is drifting downward from t_{pa} it may encounter no controlling pb swarm. In that case, the drift continues and t_{sf} may sink to t_{pc}, or even lower if no swarm at all interferes.

The value of t_{sf} may pass from one preferred temperature to another, in any order, even ignoring intermediate ones.

When the number of motes in a melt is small, the several swarms will also be small, perhaps consisting of only one mote each. In such a case the values of t_{sf} will be continually varying from one preferred temperature to another, its graph will be exceedingly jagged, long jumps and short will be intermingled, but the extreme points of the several jumps will be found to lie, in the main, near one or another of the several preferred temperatures, and nonterminal observations will probably tend to do likewise. Examples of such graphs are those of $C14$ (fig. 7) and of CI (fig. 8).

It may happen that early in the life of a specimen all the major motes of the materials for which t_p is high shall have passed the critical size (C) except some very large ones—so large that many months or years may be required for even the smallest to become, by solution at room temperature, so small as to be able to take control. In such a case t_{sf} may sink to a low value, and may remain low for a long time, only to rise rapidly thereafter when that major mote of high proper temperature takes control. Striking examples of such a change are $P12$ (fig. 9) and $C10$ (fig. 19).

There remains to be considered an effect arising from the manner in which a determination of t_{sf} has to be made. It is necessary to cool the specimen progressively from a higher temperature to that at which it freezes spontaneously. If the last is a preferred temperature, and has remained essentially unchanged, during the latter part of the cooling, no difficulty arises. However, if the temperature at which freezing occurs is not a preferred temperature, but is one that drifts with the time, one is concerned with a matching of the falling temperature of the bath with the progressively drifting value of t_{sf}. Then the value found will depend upon the rate of cooling, and will not be reproducible. One determines merely a transient value. Furthermore, other things being the same, it will take longer to obtain a match if t_{sf} is drifting toward lower temperatures than if its drift is the reverse. It will be noticed that, in conformity with this, by far the greater number of dots that are accompanied on the graphs by interrogation points, indicating that freezing occurred unusually quickly, lie each above the preceding dot; i.e., t_{sf} was presumably rising.

Examples of every one of the types of secular variation that have been considered in this section have been observed, and have been pointed out in an earlier portion of this report. Furthermore, it seems that every one of the graphs of observed values of t_{sf} (fig. 3 to 27) can by itself be accounted for by some suitable combination of these several types of variation.

h. Advantages of the New Theory

Of the eleven types of observations that have in an earlier section been compared with the demands of the older theories and have been found to conflict more or less with them, none conflict with the demands of this. (a) The paucity of singularities arises from the paucity of effective motes; (b) supercooling is to be expected; (c) a small specimen will freeze spontaneously at a higher temperature than a larger one if it contains a more efficient mote; otherwise it will not; (d) heterogeneity of a melt is to be expected; (e) nominally identical specimens should be expected to differ, each having a characteristic behavior; (f) preheating may be expected to facilitate a solution of the motes, and if they are minor ones, will thus lead to a lowering of t_{sf}, the lowering being usually permanent; (g) suitable filtration will depress the temperature of spontaneous freezing; (h) there is no reason for supposing that prolonged cooling will, of itself, affect the temperature of spontaneous freezing; (i) the number of viable embryos produced in a given volume of melt at a given temperature and in a given interval of time (τ) should be expected to increase monotonously with τ to an asymptotic maximum characteristic of the specimen (it being assumed that a mote gives rise to only a fixed, limited number of embryos, probably to but one), and the lower the temperature, the greater should be that maximum value; (j) it should be possible to initiate freezing by mechanical means; (k) preferred temperatures are to be expected.

Among other advantages of the new theory, the following may be mentioned:

1. Without any special assumption the new theory is, on its face, applicable to every melt and to every solution—it is known that molecules of solute are adsorbed at interfaces of their solutions with foreign substances.

2. The theory eliminates every demand for multiple collisions of free molecules of the melt—the molecules have already been assembled by the foreign substance.

3. It eliminates, in general, every necessity for postulating that the foreign body has a particular chemical or crystalline state. Special cases may, however, require that it shall have such a state.

4. It makes the convexity of the interface a matter of prime importance in the spontaneous freezing of the melt—whether that interface is a portion of the wall of the container, or of a foreign body placed in the melt, or of a suspended mote, is a matter of indifference.

5. It furnishes a rationale for the initiation of freezing by a rubbing of one interface against another—such rubbing may tear the adsorbed layer on the surfaces rubbed or on the motes caught between them.

6. It suggests that splashing will initiate freezing only if it is of such type and violence that an adsorbed layer is ruptured—such rupture may well attend cavitation, and may occur on the suspended motes as well as on the walls of the container. And minute particles with their adsorbed layers may be torn from the wall, and those layers may be ruptured.

7. It anticipates that a simple pouring of a supercooled melt will not initiate freezing unless it brings the melt into contact with a crystal of itself or with one isomorphous therewith, or with an adsorbed layer of effective convexity.

8. It foresees the possibility of apparently erratic results of various types, arising from various causes: from changes in the convexity of the interface, arising from solution, or precipitation, or clumping, or disruption; from the casual addition or removal of motes; from a transfer of certain motes from the bulk of the liquid to the portion of the wall that is in contact with the gas phase, and conversely; from uncontrolled factors attending the initiation of freezing by rubbing; from variations in the time that elapses between the inception of an embryo and the appearance of the resulting freezing; etc.

9. It gives no reason for expecting that freezing will be initiated by any kind of mechanical shock that does not involve either the sliding of one solid surface over another, or such a disturbance of the liquid as will cause actual or incipient cavitation, or bring the liquid into contact either with a suitable crystal, or with an adsorbed layer of efficient convexity.

10. It gives no reason for believing that freezing will be initiated by the mere squirting of a supercooled melt through a nozzle, unless there is an adsorbed layer of efficient convexity either on the walls of the nozzle or in

the air beyond it, or unless the velocity is sufficient to cause an incipient cavitation somewhere in the stream.

11. It demands that a given unchanging specimen shall freeze spontaneously at a definite temperature—above that temperature it will not freeze spontaneously; and it cannot be supercooled below that temperature.

Since all these characteristics of the theory accord with the pertinent observations reported in the earlier portions of this paper, one may say that those observations are as one would expect them to be if this new theory were the correct explanation. And the theory is of such a kind that it can be extended to other cases—to other substances and to solutions—and can be used to forecast phenomena not yet observed, such as a variation in the extent of supercooling with the strength of adsorption of the melt by the motes.

Consequently, the proposal of this theory seems to be justified. However, the author will be glad to see it replaced by a simpler, more satisfactory one that accounts as well, or better, for the experimental data herein reported.

3. EARLIER RELATED SUGGESTIONS

One will at once recall that, among the various suggestions that had previously been considered, there were several that appealed in one way or another to the adsorbed layer carried by the walls and particles. Some offered no clear idea as to how that layer achieved the assumed effect; others attributed its action to some particular state of the layer; none suggested the entirely general type of action that is appealed to in the new theory just proposed.

However, the "crystalline adsorbate" suggested by Richards [67] and the "crystal lattice" which Hammer [43] ascribed to the adsorbed layer may be satisfactory explanations in certain cases, and find their place in the new theory, though not used in the particular way suggested by the earlier authors.

Furthermore, Frankenheim [32], as early as 1860, had suggested that the initiation of crystallization by rubbing the glass-liquid interface arose from a removal of a layer of something from the glass. He thought that the rubbed portion was more readily wetted than was the adjacent portion, and attributed the crystallization to that fact.

D. SUMMARY

After an investigation of the supercooling and freezing of many specimens of water—distilled and natural, air-free and air saturated, from many sources—and a study of the effect of age, of heating, of prolonged chilling, of thermal shock, of distillation, of filtration, of solutes, and of mixing differing specimens, the results so obtained were compared with those which have been widely believed to be demanded by the kinetic theory of freezing, herein called the homogeneous theory. So many incompatibilities were found that it seemed obvious that the common understanding of that theory is wrong. The error was found to reside in certain tacit assumptions that are incompatible with the constraints inherent in the problem, but which facilitate the mathematical treatment that has been supposed to be applicable. With the elimination of those assumptions, certain phenomena that have been thought to be demanded by the theory—such as the belief that a supercooled melt is inherently unstable—are left without foundation. Furthermore, it has been found that the belief mentioned had been shown by J. W. Gibbs, nearly seventy years ago, to be incompatible, in general, with the demands of thermodynamics.

The homogeneous theory of freezing is entirely inapplicable within the range of supercooling covered by this work; and the heterogeneous theory was still too indefinite to be of much use. Consequently, a new and explicit heterogeneous theory of freezing, which accounts qualitatively for the observed phenomena, has been outlined and discussed.

The effect of each of various types of mechanical disturbances upon the freezing of supercooled water has been studied and so has that of each of several solutes upon the initiation of freezing by the rubbing together of immersed solids.

Incidental observations on the growth of spicules of ice, the propagation of freezing along the wall of the container above the meniscus, the growth of ice by reverse sublimation and through the water, adhesion of ice, quiet boiling, remarks on the bursting of water pipes by freezing, etc. have been recorded.

In conclusion I wish to express my sincere appreciation to the National Bureau of Standards and to its present Director, Dr. E. U. Condon, for extending to me facilities for this work, even after my retirement, and to its Associate Director, Dr. E. C. Crittenden, for arranging for the publishing of this report; to a number of my former colleagues, including Dr. D. N. Craig, Mr. F. W. Schwab, Dr. F. D. Rossini, Dr. E. Wichers, Mr. W. Spangenberg, Dr. C. P. Saylor, Mr. E. F. Mueller, Mr. A. I. Dahl, for assistance of various kinds; to Dr. L. J. Briggs, Dr. H. L. Dryden, and Mr. W. F. Stutz, for laboratory space during war-time congestion; and especially to the then Director of the Bureau, Dr. Lyman J. Briggs, for his unfailing interest and encouragement throughout the entire course of the work.

APPENDIX

SEALED BULBS: CONTENTS AND HISTORIES

All "C" bulbs are of soft soda glass; only the first 4 are marked with a C, the others carry the numeral only.

The "G" bulbs are all of Pyrex, and are each marked with a *green* arabic numeral. They do *not* carry a "G." They were prepared by Mr. F. W. Schwab.

The "P" bulbs are all of Pyrex, and each carries a P and an arabic numeral.

In general a bulb has been frozen frequently during a period immediately following its sealing, and at longer intervals later.

CI. Charged by vacuum distillation without ebullition; air removed from the system by prolonged boiling. Sealed August 31, 1936. Freezings before January 8, 1937, were made in a test tube containing alcohol, immersed in a cold bath in a Thermos bottle. The cold bath was a mixture of ice, salt, and water, until December 7, 1936; then alcohol cooled by means of solid CO_2, until January 7, 1937. After that, the bulb was suspended in alcohol contained in an unsilvered Dewar cylinder, and cooled by means of solid CO_2.

The sealed bulb was heated in boiling water for a short time on October 21, 1936; for ½ hour on May 15, 1937; for ¼ hour, and later for ½ hour on May 17, 1937; for 2 hours on May 18, 1937. Tables 7, 11; fig. 8.

CII. Charged as was CI. Sealed September 15, 1936; during the annealing, a short fine crack, thought to be of no significance, developed in the seal. Early observations in which an ordinary Thermos bottle was used, indicated that in September of 1936 t_{sf} lay between − 9 and −14°C, in October and November it was below − 14°C, on December 9 it was − 18°C. The next freezing was on January 5, 1937, and in an unsilvered Dewar cylinder, as used thereafter; freezing was seen to proceed slowly from a single small rosette, a phenomenon that was later found to characterize all water-alcohol mixtures. From January 5 to 11 t_{sf} was − 22.2°C. On January 11, 1937, the top of the stem cracked off; it was resealed, but imperfectly. Immediately thereafter t_{sf} was −25.3°C (not shown in fig. 27), and alcohol from the bath was seen to be entering through the imperfect seal; that ice was in equilibrium with the melt at about −6.5°C, indicating a concentration around 14 per cent of alcohol by weight. No more freezings were made. On January 28, 1937, CII was opened

and its contents were poured into C18, which was then sealed; CII was destroyed. (A significant amount of alcohol may have evaporated between January 11 and the sealing of C18 on January 28.) Fig. 27.

CIII. Like CI. Sealed November 13, 1936. Cooling bath as for CI and CII. Tables 3, 7; fig. 5.

CIV. Like CI except that air was admitted before the bulb was sealed off from the reservoir. Sealed November 16, 1936. Table 7; fig. 22.

C5, C6. Destroyed. Not used.

C7. Contains some of the residue from which CIV had been distilled. Atmospheric pressure. Sealed November 16, 1936. Tables 7, 10, 11; fig. 22.

C8. Distilled water from stock bottle; atmospheric pressure. Sealed November 27, 1936. Tables 7, 11; fig. 3.

C9. Similar to C8. Sealed November 27, 1936. Table 7; fig. 3.

C10. Conductivity water; charged by Dr. Craig; atmospheric pressure. Specific conductivity, 0.084 × 10^{-6} (ohm-cm)$^{-1}$ at 23°C. Sealed December 10, 1936. Tables 2, 3, 7; fig. 19.

C11. Similar to C10; charged and sealed at the same time. Heated in boiling water for 3 minutes on December 11, 1936, and for 17¾ hours on January 15 to 18, 1937; kept continuously between − 8.0 and − 10.3°C from January 23, 1938, to December 2, 1938 (312 days), and during the immediately preceding 41 days its temperature had lain at or below − 6°C except for a brief period at + 3°C. Tables 6, 7; fig. 19.

C12. Pt wire sealed through wall slightly above the meniscus; charged from a fresh stock of distilled water; atmospheric pressure. Sealed December 21, 1936. Heated in boiling water for 18½ hours on January 5–7, 1937; 17⅔ hours on January 15–18. On February 26 cut top from stem and left the so-opened bulb in book case for 5 days and 1 hour; resealed March 3, 1937. Tables 2, 3, 7; fig. 15.

C13. Charged from the same water as was C12; water in bulb is covered with castor oil. Sealed December 21, 1936; atmospheric pressure. Heated to 54°C for 8 hours on March 26, 1937; in boiling water for 2 hours on March 29, for 2 hours on March 31, and for 5½ hours on April 8, 1937. Table 7; fig. 14.

C14. Vacuum distilled, without ebullition, from water acidulated with H_2SO_4 and $K_2Cr_2O_7$. Air removed by prolonged boiling. Low pressure. Sealed December 22, 1936. Tables 5, 7; fig. 7.

318

C15. Fresh snow-water; from midlayer of melt; atmospheric pressure. Sealed February 17, 1937. Table 7; fig. 9.

C16. Otherwise used.

C17. Ordinary distilled water; atmospheric pressure. Sealed December 30, 1936. The stem was heated on January 8, 1937, so that water poured into it was suddenly and completely vaporized, leaving solute on the walls. This was done several times. Tables 7, 11; fig. 6.

C18. Alcohol-water mixture from CII; atmospheric pressure. Sealed January 28, 1937. Tables 10, 11; fig. 27.

C19. Distilled water from stock bottle; atmospheric pressure. Sealed January 29, 1937. Table 7; figs. 6, 27.

C20. Distilled water (from same lot as C19) and alcohol; 19.8 per cent alcohol by weight; atmospheric pressure. Sealed January 29, 1937. Fig. 27.

C21. Exactly similar to C20, except that the mixture is 5.6 per cent alcohol by weight; atmospheric pressure. Sealed January 29, 1937. Fig. 27.

C22. Cracked. Discarded.

C23. Charged from same snow-water as is C15, but this water was poured into the bulb; atmospheric pressure. Sealed February 17, 1937. Tables 3, 7, 10; fig. 4.

C24. Similar to C23 but contains less of the scum that formed on the melted snow; atmospheric pressure. Sealed February 17, 1937. Table 7; fig. 10.

C25. Water from the large spring at "Magnolia," a farm in Maryland, about 22 miles from Washington; atmospheric pressure. Sealed March 1, 1937. Table 7; fig. 4.

C26. Water and green algae from the "old" spring at "Magnolia"; atmospheric pressure. Sealed March 1, 1937. Table 7; fig. 11.

C27. Water and sediment from a pool in the marshy ground adjacent to the "old" spring; atmospheric pressure. Sealed March 1, 1937. Table 7; fig. 3.

C28. Water from the Little Patuxent River. The water was dipped from the river, and the top layer was decanted before the water was poured into the bulb; atmospheric pressure. Sealed March 1, 1937. Table 7; fig. 10.

C29. Water from the cold-water faucet at "Magnolia"; atmospheric pressure. Sealed March 1, 1937. Table 7; fig. 13.

C30. Water from the hot-water faucet at "Magnolia"; atmospheric pressure. Sealed March 1, 1937. Table 7; fig. 13.

C31. Water taken directly from the pump at "Magnolia"; atmospheric pressure. Sealed March 1, 1937. Table 7; fig. 13.

C32. Snow-water that had been bottled by another in the winter of 1935–1936; atmospheric pressure. Sealed March 1, 1937. Table 7; fig. 11.

C33. Water from the hot-water faucet in the laboratory; no water had been drawn from that faucet for a week or more. The water was about room temperature; atmospheric pressure. Sealed March 3, 1937. Table 7; fig. 3.

C34. Water taken from the same faucet as for C33, after water had been run off until it came out fairly hot; atmospheric pressure. Sealed March 3, 1937. Table 7; fig. 3.

C35. Water from the cold-water faucet in the laboratory; much water had previously been drawn that day. Atmospheric pressure. Sealed March 3, 1937. Warmed to 54°C for 8 hours March 26, 1937. Heated in boiling water for 2 hours March 29, 2 hours March 31, 5½ hours April 8, 1937. Kept continuously between − 8.0 and − 10.3°C January 23, 1938, to December 2, 1938 (312 days), and during the immediately preceding 41 days its temperature had lain at or below − 6°C except for a brief period at + 3°C. Tables 6, 7; fig. 14.

C36. Destroyed. Never used.

C37. Charged with residue from the boiler used for steaming the bulbs. Atmospheric pressure. Sealed March 4, 1937. Heated to 54 °C for 8 hours on March 26; in boiling water for 2 hours on March 29, for 2 hours on March 31, for 5½ hours on April 8, 1937. Table 7; fig. 16.

C38. Water from a clear, swift brook south of Tilden St. Atmospheric pressure. Sealed March 4, 1937. Tables 3, 7; fig. 26.

C39. Water from below the surface of a scummed-over, stagnant pool south of Tilden St. Atmospheric pressure. Sealed March 4, 1937. Table 7; fig. 26.

C40. Differs from C39 only in that the water was dipped from the surface of the pool. Atmospheric pressure. Sealed March 4, 1937. Tables 3, 7; fig. 4.

C41. Water from the bottom of the same pool as that from which contents of C39 and C40 were taken. The charge contains some of the bottom material of the pool. Atmospheric pressure. Sealed March 4, 1937. Table 7; fig. 26.

C42. Water from an aquarium, taken distant from the walls and where the water was clear. Atmospheric pressure. Sealed March 4, 1937. Tables 2, 7; fig. 12.

C43. Differs from C42 only in that the water was taken from the alga-covered walls and from leaves and stems of the sea-weed growing in the aquarium. Atmospheric pressure. Sealed March 4, 1937. Table 7; fig. 11.

C44. Destroyed. Not used.

*C*45. Melt of a clear crystal of artificial ice. Atmospheric pressure. Sealed March 6, 1937. Heated to 54°C for 8 hours on March 26; in boiling water for 2 hours on March 29, for 2 hours on March 31, for 5½ hours on April 8, 1937. Table 7; fig. 16.

*C*46. Water from the cold-water faucet of the laboratory (*cf.* *C*35); in the water are long pieces of 2-mm copper wire. The tops of the wires project above the meniscus. Atmospheric pressure. Sealed March 13, 1937. Table 7; fig. 26.

*C*47. Differs from *C*46 only in that the pieces of wire are shorter, and are completely submerged. Atmospheric pressure. Sealed March 13, 1937. Table 7; fig. 10.

*C*48. Differs from *C*46 and *C*47 only in that it contains no copper wire. Atmospheric pressure. Sealed March 13, 1937. Table 7; fig. 3.

*C*49. Distilled water from a small wash-bottle which had been undisturbed for 4 months. Atmospheric pressure. Sealed November 15, 1937. The water was kept frozen continuously for 4 hours on December 6, and again on December 7, 1937. The specimen was kept continuously between − 8.0 and − 10.3°C January 23, 1938, to December 2, 1938 (312 days); it froze and was melted 4 times prior to February 12, after which it did not freeze. Tables 6, 7, 11; fig. 12.

*C*50. The bulb was cleaned, filled with distilled water, immersed in a beaker of the same water and boiled for 6¼ hours. It was then charged with condensate that had been collected from a Pyrex still which was charged with distilled water that had been boiled with reflux condenser for 15 hours. Done in dusty room. Atmospheric pressure. Sealed July 27, 1943. Table 7; fig. 20.

*C*51. Bulb was treated as was *C*50. Charged directly from the condenser of a Pyrex still that was charged with the distillate from water that was taken directly from the pump at "Magnolia" (see also *C*31). Done in dusty room. Atmospheric pressure. Sealed July 29, 1943. Heated in boiling water for 5½ hours on September 14–15, 1943. Figs. 16, 21.

*G*1. Prepared, charged, and sealed by Mr. Schwab. The charge was taken directly from an all-quartz still. Received sealed on July 5, 1943. Atmospheric pressure. Heated in boiling water for 9 hours on September 22, for 14 hours on September 27–28, for 26½ hours on October 8–13, 1943. Tables 2, 7; fig. 18.

*G*2. Like *G*1 in every way. Heated like *G*1 and at the same times. Table 7; fig. 18.

*G*3. Like *G*1 in every way. Heated like *G*1 and at the same times. Tables 7, 10, 11; fig. 18.

*G*4. Like *G*1 in every way. Heated like *G*1 and at the same times. Table 7; fig. 18.

*G*5. Like *G*1 in every way. Heated like *G*1 and at the same times. Tables 2, 3, 7; fig. 18.

*P*1. Conductivity water. Similar to *C*10 and *C*11; charged at the same time. Sealed December 10, 1936. Table 7; fig. 19.

*P*2. A duplicate of *P*1. Atmospheric pressure. Sealed December 10, 1936. Heated in boiling water for 3 minutes on December 11, 1936, for 17¾ hours on January 15–18, 1937. Tables 3, 7; fig. 19.

*P*3. Destroyed. Not used.

*P*4. Water from below the surface of a sample of water taken March 4, 1937, from swampy ground south of Tilden St. and kept at 0°C until March 8, and at room temperatures thereafter. Atmospheric pressure. Sealed March 13, 1937. Table 7; fig. 11.

*P*5. Differs from *P*4 only in that it contains a long, but completely submerged, piece of 2-mm copper wire. Atmospheric pressure. Sealed March 13, 1937. Table 7; fig. 10.

*P*6. Mixture of equal volumes of the water used for *P*4 and *P*5 and of water from the cold-water faucet in the laboratory (*cf.* *C*35, *C*46, *C*47, *C*48). Atmospheric pressure. Sealed March 13, 1937. Table 7; fig. 11.

*P*7. Melted snow which had been in a bookcase since February 17, 1937, when some of it was placed in *C*15. Atmospheric pressure. Sealed March 25, 1937. Heated in boiling water for 2 hours on March 31, for 5½ hours on April 8, 1937. Table 7; fig. 15.

*P*8. Water from the cold-water faucet in the laboratory (*cf.* *C*35, *C*46–48, *P*6). Atmospheric pressure. Sealed March 26, 1937. Heated in boiling water for 2 hours on March 31, for 5½ hours on April 8, 1937. Table 7; fig. 4.

*P*9. Set aside for another purpose.

*P*10. Water that was vacuum distilled, without ebullition, and condensed in *P*10. Distilled from midlayers of old stock of distilled water; air removed by pump. Sealed May 20, 1937. Heated in boiling water 2¼ hours June 2, and 4¼ hours June 4, 1937. Kept continuously between − 8.0 and − 10.3°C from January 23, 1938, to December 2, 1938 (312 days), and during the immediately preceding 41 days its temperature had lain at or below − 6°C, except for a brief period at + 3°C. Tables 3, 6, 7; fig. 5, 24.

*P*11. Contains the residue from the distillation of charges into *P*10 and *P*12. Pump exhausted. Sealed May 21, 1937. Heated in boiling water for 2¼ hours on June 2, for 4¼ hours on June 4, 1937. Table 7; figs. 16, 24.

*P*12. Similar to *P*10. Pump exhausted. Sealed May 21, 1937. Heated in boiling water for 2¼ hours on June 2, for 4¼ hours on June 4, 1937. Table 7; figs. 9, 24.

P13. Contains a mixture of alcohol and water, and a glass bead. Atmospheric pressure. Sealed May 21, 1937. Fig. 27.

P14. Contents were distilled, under pump exhaustion and without ebullition, from the residue from which the contents of *P16* was distilled. Low pressure. Sealed May 28, 1937. Heated in boiling water for 2¼ hours on June 2, for 4¼ hours on June 4, 1937. Table 7; figs. 6, 25.

P15. Contains residue from low pressure, non-ebullition distillation into *P16* and *P14.* Low pressure (pump exhausted). Sealed May 28, 1937. Heated in boiling water for 2¼ hours on June 2, for 4¼ hours on June 4, 1937. Table 7; fig. 25.

P16. Contents were distilled, under pump exhaustion and without ebullition, from water from midlayers of old stock of distilled water. Low pressure. Sealed May 26, 1937. Heated in boiling water for 2¼ hours on June 2, for 4¼ hours on June 4, 1937. Table 7; figs. 15, 25.

P17. Contents were distilled, under pump exhaustion and without ebullition, from water taken from the midlayers of old stock of distilled water. Low pressure. Sealed May 29, 1937. Table 7; fig. 23.

P18. Contains the residue from the distillation into *P17.* Pump exhausted. Sealed May 29, 1937. Table 7; figs. 9, 23.

P19. Used for another purpose.

P20. Charged with water taken from below the surface of water that contained glass beads and bits of copper wire and that had been used two years before for steaming bulbs; it had been undisturbed in a corked flask since July 1937. A glass bead was inadvertently lifted with the water and placed in *P20.* (Hence the water came from near the bottom of the flask.) Atmospheric pressure. Sealed June 23, 1939. It was so sealed that the stem was long. On May 1, 1943, in the dusty room, the upper end of the stem was sealed off; during the process a hole, sealed promptly, was blown in the stem. Table 7; fig. 12.

P21. Contents vacuum distilled from distilled water filtered through a German "19*G*" fritted-glass filter; in the dusty room. Air removed by prolonged boiling. Vacuum. Sealed April 29, 1943. Table 7; fig. 20.

P22. Bulb cleaned and given to Mr. F. W. Schwab for recleaning and charging directly from the Bureau still. He returned it to me, charged and sealed at atmospheric pressure on July 5, 1943. Heated in boiling water for 9 hours on September 22, for 14 hours on September 27–28, for 26½ hours on October 8–13, 1943. Table 7; figs. 6, 17.

P23. Like *P22* in every way except heating. Heated in boiling water 3 hours September 14 to 15, 1943, 14 hours September 27 to 28, and 26½ hours October 8 to 13, 1943. Table 7; fig. 17.

P24. Like *P22* in every way, including its heating. Tables 2, 7; fig. 17.

P25. Like *P23* in every way, including its heating. Tables 2, 3, 7; fig. 17.

P26. Like *P22* in every way, including its heating. Tables 2, 7; fig. 9.

P27. Like *P22* in every way, including its heating. Table 7; fig. 17.

P28. Filled bulb with distilled water, immersed in beaker of same water, and boiled 6¼ hours. Emptied, rinsed, and charged with distillate from distilled water that had been boiled with reflux condenser (Pyrex) for 15 hours. All done in the dusty room. Atmospheric pressure. Sealed July 27, 1943. Table 7; fig. 20.

P29. Charged with water that was directly distilled into it in a vacuum and without ebullition. The water had previously been filtered through a "19*G*" fritted-glass filter and brought to boiling. Air removed by prolonged boiling, outlet of system being immersed in boiling water; all similar to *P21.* All done in the dusty room. Vacuum. Sealed April 30, 1943. Table 7; fig. 20.

P30. Charged with some of the boiling water that was later used for charging the system for charging *P29*; in dusty room. Atmospheric pressure. Sealed April 30, 1943. Table 7; figs. 3, 20.

P31. Charged by vacuum distillation without ebullition, air being removed by boiling. The water had been passed through a fine ("*F*") Pyrex fritted-glass filter before the distillation. All done in the dusty room. Vacuum. Sealed on May 6, 1943. Table 7; fig. 20.

P32. Charged by low-pressure distillation, without ebullition, from what was left in the reservoir after the charging of *P31*; the low pressure was that supplied by the laboratory low-pressure line—some 10 to 20 cm-Hg. All done in the dusty room. Low-pressure. Sealed on May 31, 1943. (In the process of sealing there was a blow-out which was quickly sealed with fused glass; the entire bulb was hotter than could be held in the hands.) Table 7; fig. 20.

P33. Charged with water taken directly from the condenser of a Pyrex still which was charged with ordinary distilled water that had been boiled with a reflux Pyrex condenser for 15 hours; in dusty room. Atmospheric pressure. Sealed on July 20, 1943. Table 7; figs. 3, 20.

P34. Charged in the same way as *P33*, and shortly before it. The only noted difference had to do with the window directly back of the still. That was open until *P34* was ready for sealing; it was then closed, and remained closed until after the sealing of *P33*. All done in the dusty room. Atmospheric pressure. Sealed on July 20, 1943. Table 7; fig. 20.

P35. The bulb filled with ordinary distilled water was immersed in a beaker of the same water, and was boiled for 6¼ hours. Then it was removed, emptied, rinsed with distilled water, and inverted to drain. In a dust-proof box it was rinsed and then charged with water brought from the pump at "Magnolia." All done in the dusty room. Atmospheric pressure. Sealed on July 27, 1943. Figs. 16, 21.

P36. Prior to charging, the bulb was treated as were *C*50, *C*51, *P*28, and *P*35, and then placed inverted in an apparatus case. Charged with water taken directly from the condenser of a Pyrex still charged with water brought from the pump at "Magnolia." All done in the dusty room. Atmospheric pressure. Sealed on July 28, 1943. Fig. 21.

P37. Charged with the boiled, triply distilled water from the water brought from the pump at "Magnolia"; in dusty room. Atmospheric pressure. Sealed September 22, 1943. Fig. 21.

REFERENCES

When a journal has been issued in a number of series, each separately numbered, the designation of the series, or of its analogues, enclosed in parentheses, has been placed before the number of the volume. For example, see No. 5.

The Royal Society of Canada is split into several sections, and each section publishes Transactions, independent of the other sections, and those Transactions are issued in successive series of independently numbered volumes. In this list, the number of the section of the society is indicated by a roman numeral; that is followed by the number of the series, indicated by an arabic numeral enclosed in parentheses; and then comes the number of the volume in bold face. For example, see No. 37.

In some cases articles are reissued in the form of a series of reprints, the series carrying a distinctive designation and the article a distinctive number. In such cases the complete series designation has been placed in parentheses immediately after the page number. For example, No. 28. In such cases, time is usually saved by ordering the article by its series designation. For that reason, it has seemed worth while to give that designation.

In the case of Gibbs (38) and of Graham (42) reference is made both to the journal article and to the collected papers. The latter may be more readily accessible than the former.

1. ALTBERG, W. J. 1938. Crystallization nuclei in water. *Acta phys.-chim. U. R. S. S.* **8**: 677–678.
2. —— 1938. Wasserkristallisationsversuche. I. *Acta phys.-chim. U. R. S. S.* **9**: 725–732.
3. ALTBERG, W., and W. LAVROW. 1939. Experiments on the crystallization of water. II. *Acta phys.-chim. U. R. S. S.* **11**: 287–290.
4. ALTBERG, W., and W. LAVROW. 1940. Experiments on the crystallization of water. III. *Acta phys.-chim. U. R. S. S.* **13**: 725–729.
5. ALTY, T. 1933. The maximum rate of evaporation of water. *Philos. Mag.* (7) **15**: 82–103.
6. AVRAMI, M. 1939. Kinetics of phase change. I. General theory. *Jour. Chem. Phys.* **7**: 1103–1112.
7. —— 1941. Granulation, phase change, and microstructure. Kinetics of phase change. III. *Jour. Chem. Phys.* **9**: 177–184.
8. BALLY, O. 1935. Ueber eine eigenartige Eiskrystallbildung. *Helv. chim. Acta* **18**: 475–476.
9. BARNES, H. T. 1906. Ice formation. N. Y., Wiley.
10. —— 1925. Colloidal water and ice. Coll. symp. monog. III: 103–111.
11. —— 1926. Colloidal forms of water and ice. *In* Colloid Chemistry (J. Alexander, ed.) I: 435–443.
12. —— 1929. The science of ice engineering. *Sci. Mon.* **29**: 289–297.
13. BARNES, T. C., and T. L. JAHN. 1934. Properties of water of biological interest. *Quart. Rev. Biol.* **9**: 292–341.
14. BERGMANN, P. G. 1943. The formation of centers of condensation in supercooled phases. *Phys. Rev.* (2) **63**: 456.
15. BERKELEY, EARL OF. 1912. Solubility and supersolubility from the osmotic standpoint. *Philos. Mag.* (6) **24**: 254–268.
16. BIILMANN, E., and A. KLITT. 1932. Untersuchungen über Kristallkernbildung bei Piperonal und Allozimtsäure. *K. danske vidensk.* (Math.-fys.-medd.) 12 (4): 51 pages.
17. BLOCH, R., T. BRINGS, and W. KUHN. 1931. Ueber das Problem der Ueberhitzbarkeit von Kristallkeimen. *Ztschr. phys. Chem.* (B) **12**: 415–426.
18. BOROVIK-ROMANOVA, T. 1924. Undercooling of water in capillaries. (Russian) *Jour. Soc. phys.-chim. russe* (Phys.) **56**: 14–22. (See *Chem. Abst.* **19**: 3186, 1925.)

19. BROWN, F. C. 1916. The frequent bursting of hot water pipes in household plumbing system. *Phys. Rev.* (2) **8**: 500–503.
20. BUERGER, M. J. 1936. The kinetic basis of crystal polymorphism. *Proc. Natl. Acad. Sci.* **22**: 682–685.
21. DE COPPET, L. C. 1872. Sur la température de cristallisation spontanée de la solution sursaturée de sulfate de soude. *Bull. Soc. chim. Paris* (n.s.) **17**: 146–155.
22. —— 1875. Théorie de la surfusion et de la sursaturation, d'après les principes de la théorie mécanique de la chaleur. *Ann. Chim. et Phys.* (5) **6**: 275–288.
23. —— 1907. Recherches sur la surfusion et la sursaturation. *Ann. Chim. et Phys.* (8) **10**: 457–527.
24. DANILOW, V., and W. NEUMARK. 1937. Ueber das Vorhandensein von Kristallisationskeimen oberhalb des Schmelzpunktes und die Struktur der Flüssigkeiten. *Phys. Ztschr. Sowjet.* **12**: 313–323.
25. DEHLINGER, U., and E. WERTZ. 1941. Keimbildung in wässeriger Lösung. *Ann. Phys., Lps.* (5) **39**: 226–240.
26. DESPRETZ, C. 1837. Recherches sur le maximum de densité des liquides. *C. R. Acad. Sci., Paris* **4**: 124–130.
27. —— 1839. Recherches sur le maximum de densité de l'eau pure et des dissolutions aqueuses. *Ann. Chim. et Phys.* (2) **70**: 5–81.
27a. DORSEY, H. G. 1921. Peculiar ice formation. *Phys. Rev.* (2) **18**: 162–164.
28. DORSEY, N. E. 1938. Supercooling and freezing of water. *Jour. Res. Natl. Bur. Stand.* **20**: 799–808 (RP-1105).
29. —— 1940. Properties of the ordinary water-substance. (*Amer. Chem. Soc. Monog.* No. 81.) N. Y., Reinhold.
30. DUFOUR, L. 1861. Sur la congélation de l'eau et sur la formation de la grêle. *Arch. Sci. phys. nat.* (n.p.) **10**: 346–371. See also **11**: 22–30. Rev. in *Ann. Phys., Lps.* (Pogg.) **114**: 530–554.
31. ERLENMEYER, H. 1930. Bemerkungen über die Trachten gekletterter Krystalle. *Helv. chim. Acta* **13**: 1006–1008.
31a. FRANK, F. C. 1946. Molecular structure of deeply supercooled water. *Nature* **157**: 276.
32. FRANKENHEIM, L. 1860. Ueber das Entstehen und das Wachsen der Krystalle nach mikroskopischen Beobachtungen. *Ann. Phys., Lps.* (Pogg.) **111**: 1–60.
33. FRENKEL, J. 1935. The liquid state and the theory of fusion. II. The theory of fusion and crystallization. *Acta phys.-chim. U. R. S. S.* **3**: 913–938.
34. —— 1939. A general theory of heterophase fluctuations and pretransition phenomena. *Jour. Chem. Phys.* **7**: 538–547.
35. FRICKE, R. 1934. Zur Ausscheidung von Kristallen und Gasen aus übersättigten Lösungen. *Kolloidztschr.* **68**: 165–168.
36. FÜCHTBAUER, C. 1904. Die freiwillige Erstarrung unterkühlter Flüssigkeiten. *Ztschr. phys. Chem.* **48**: 549–568.
37. GARRARD, J. D. 1924. The technique of preparation of dust-free liquids by distillation. *Trans. Roy. Soc. Can.* III (3) **18**: 126–127.
38. GIBBS, J. W. 1878. On the equilibrium of heterogeneous substances. *Trans. Conn. Acad. Arts Sci.* **3**: 108–248, 343–524. Or, Scientific Papers I: 55–353. N. Y., Longmans, Green. 1906. (Page references in the text are to the Papers.)
39. GOETZ, A. 1935. Experimental evidences of group phenomena in the solid metallic state. *Int. Conf. Phys., London, 1934* **2**: 62–71. Cambridge Univ. Press.
40. —— 1935. The phenomenon of supercooling. *Phys. Rev.* (2) **47**: 257.

41. GOPAL, R. 1943. Supersaturation limits of solutions. I. Jour. Indian Chem. Soc. 20: 183–188.
42. GRAHAM, T. 1831. On the influence of air in determining the crystallization of saline solutions. Trans. Roy. Soc. Edinburgh 11: 114–118. Or, Chemical and physical researches. Edinburgh Univ. Press. 1876.
43. HAMMER, C. 1938. Untersuchungen der Kristallkeime. Ann. Phys., Lpz. (5) 33: 445–458.
44. HINSHELWOOD, C. N., and H. HARTLEY. 1922. The probability of spontaneous crystallization of supercooled liquids. Philos. Mag. (6) 43: 78–94.
45. ISAAC, FLORENCE. 1908. The temperatures of spontaneous crystallisation of mixed solutions and their determination by means of the index of refraction. Mixtures of solutions of sodium nitrate and lead nitrate. Jour. Chem. Soc. 93: 384–411.
46. JAFFÉ, G. 1903. Studien an übersättigten Lösungen. Ztschr. phys. Chem. 43: 565–594.
47. KAISCHEW, R. 1937. Ueber die Möglichkeit der Bildung von Kriställchen in Schmelzen bei Temperaturen oberhalb des Schmelzpunktes. Ann. Phys., Lpz. (5) 30: 184–192.
48. KENRICK, F. B. 1922. The scattering of light: Note on Wolski's paper on optically empty liquids. Jour. phys. Chem. 26: 72–74.
49. KORNFELD, GERTRUD. 1916. Ein Beitrag zur Frage der Ueberschreitungserscheinungen. Mh. Chem. 37: 609–633.
50. LEHMANN, O. 1900. Structur, System und magnetisches Verhalten flüssiger Krystalle und deren Mischbarkeit mit festen. Ann. Phys., Lpz. (4) 2: 649–705.
51. LINK, H. F. 1839. Ueber die erste Entstehung der Krystalle. Ann. Phys., Lps. (Pogg.) 46: 258–264.
52. MARCELIN, R. 1909. Observations sur la cristallisation spontanée. C. R. Acad. Sci., Paris 148: 631–633.
53. MARTIN, W. H. 1913. The Tyndall effect in liquids. Trans. Roy. Soc. Can. III (3) 7: 219–220.
54. —— 1920. The scattering of light by dust-free liquids. Jour. Phys. Chem. 24: 478–492.
55. MASCART, E. 1896. Leçons sur l'électricité et le magnétisme, I: 249–254. Paris, Gauthier-Villars et Fils.
56. MEYER, J., and W. PFAFF. 1934. Zur Kenntnis der Kristallisation von Schmelzen. Ztschr. anorg. Chem. 217: 257–271.
57. MEYER, J., and W. PFAFF. 1935. Zur Kenntnis der Kristallisation von Schmelzen, II. Ztschr. anorg. Chem. 222: 382–388.
58. MEYER, J., and W. PFAFF. 1935. Zur Kenntnis der Kristallisation von Schmelzen. III. Ztschr. anorg. Chem. 224: 305–314.
59. MIKHNEVICH, G. L., and I. F. BROWKO. 1938. Stability of the crystallisation centres of an organic liquid at various temperatures and conclusions to be drawn therefrom concerning Tammann's method. Phys. Ztschr. Sowjet. 13: 113–122.
60. OLTRAMARE, G. 1879. Notice sur la constitution des nuages et la formation de la grêle. Arch. Sci. phys. nat. (3) 1: 487–501.
61. OSTWALD, W. 1891. Lehrbuch der allgemeinen Chemie, ed. 2, I: 1039 et seq. Leipzig, Engelmann.
62. —— 1896-1902. Lehrbuch der allgemeinen Chemie, ed. 2, II₂: 379 et seq., 776, 784. Leipzig, Engelmann.
63. OTHMER, P. 1915. Studien über das spontane Kristallisationsvermögen. Ztschr. anorg. Chem. 91: 209–247.
64. PAWLOW, P. 1909. Ueber die Abhängigkeit des Schmelzpunktes von der Oberflächenenergie eines festen Körpers. Ztschr. phys. Chem. 65: 1–35, 545–548.
65. PENNYCUICK, S. W., and C. E. WOOLCOCK. 1939. Some observations on the colloidal impurities in distilled water. Jour. Phys. Chem. 43: 681–685.

65a. RAU, W. 1944. Gefriervorgänge des Wassers bei tiefen Temperaturen. (Vorläufige Mitteilung.) Schriften dtsch. Akad. Luftfahrtforsch. 8 (2): 65–84.
66. RICE, F. O. 1926. The catalytic activity of dust particles. Jour. Amer. Chem. Soc. 48: 2099–2113.
67. RICHARDS, W. T. 1932. The persistence and development of crystal nuclei above the melting temperature. Jour. Amer. Chem. Soc. 54: 479–495.
68. RICHARDS, W. T., E. C. KIRKPATRICK, and C. E. HUTZ. 1936. Further observations concerning the crystallization of undercooled liquids. Jour. Amer. Chem. Soc. 58: 2243–2248.
69. RÖNTGEN, W. C. 1892. Ueber die Constitution des flüssigen Wassers. Ann. Phys., Lpz. (Wied.) 45: 91–97.
70. SCHAUM, K. 1898. Ueber die Krystallisation des unterkühlten Benzophenons. Ztschr. phys. Chem. 25: 722–728.
71. —— 1934. Bildung und Umwandlung von Kristallen. Ztschr. angew. Chem. 47: 110.
72. —— 1939. Studien zur Interferometrie. IV. Versuche zum Nachweis von Alterungserscheinungen an Schmelzflüssen. Ztschr. wiss. Photogr. 38: 113–125.
73. SCHAUM, K., and E. RIFFERT. 1922. Zur Kenntnis der Aggregatzustandsänderungen und des Polymorphismus. I. Ueber Kristallisation aus unterkühlten Schmelzen. Ztschr. anorg. Chem. 120: 241–260.
74. SCHAUM, K., and F. SCHOENBECK. 1902. Unterkühlung und Krystallisation von Schmelzflüssen polymorpher Stoffe. Ann. Phys., Lpz. (4) 8: 652–662.
75. SPANGENBERG, K. 1935. Wachstum und Auflösung der Krystalle. Handwörterbuch der Naturwissenschaften, ed. 2, X: 362–401. Jena, Fischer.
76. STRANSKI, I. N., and R. KAISCHEW. 1935. Kristallwachstum und Kristallkeimbildung. Phys. Ztschr. 36: 393–403.
77. TAMMANN, G. 1897. Ueber die Grenzen des festen Zustandes. Ann. Phys., Lps. (Wied.) 62: 280–299.
78. —— 1898. Ueber die Abhängigkeit der Zahl der Kerne, welche sich in verschiedenen unterkühlten Flüssigkeiten bilden, von der Temperatur. Ztschr. phys. Chem. 25: 441–478.
79. —— 1903. Kristallisieren und Schmelzen. Leipzig, Barth.
80. —— 1913. Die Beziehungen der Volumfläche zum Polymorphismus des Wassers. Ztschr. phys. Chem. 84: 293–312.
81. —— 1925. States of aggregation. (Trans. by R. F. Mehl.) N. Y., Van Nostrand.
82. —— 1926. Zur Kenntnis der molekularen Zusammensetzung des Wassers. Ztschr. anorg. Chem. 158: 1–16.
83. —— 1929. Die Abhängigkeit der Zahl der Kristallisationszentren von der Temperatur. Ztschr. anorg. Chem. 181: 408–416.
84. TAMMANN, G., and A. BÜCHNER. 1935. Die Unterkühlungsfähigkeit des Wassers und die lineare Kristallisationsgeschwindigkeit des Eises in wässerigen Lösungen. Ztschr. anorg. Chem. 222: 371–381.
85. TAMMANN, G., and H. E. VON GRONOW. 1931. Ueber die spontane Kristallisation unterkühlter Schmelzen und übersättigter Lösungen. Ztschr. anorg. Chem. 200: 57–73.
86. TIMMERMANS, J., J. DE ROOSTER, and J. MICHEL. 1939. La cristallisation spontanée et la vitesse de cristallisation de l'eau et de l'eau lourde. C. R. Acad. Sci., Paris 208: 282–283.
87. VOGELSANG, H. 1875. Die Krystalliten. Bonn, Cohen & Sohn.
88. VOLMER, M. 1929. Ueber Keimbildung und Keimwirkung als Spezialfälle der heterogenen Katalyse. Ztschr. Elektrochem. 35: 555–561.
89. VOLMER, M., and A. WEBER. 1926. Keimbildung in übersättigten Gebilden. Ztschr. phys. Chem. 119: 277–301.

90. WALTON, J. H., and A. BRAUN. 1916. The effect of dissolved substances on the velocity of crystallisation of water. *Jour. Amer. Chem. Soc.* **38**: 317–330.

91. v. WARTENBERG, H. 1939. Grenzhautbildung zwischen heisser Flüssigkeit und kalter Wand. *Ztschr. Elektrochem.* **45**: 497–502.

92. WEBSTER, W. L. 1933. Phenomena occurring in the melting of metals. *Proc. Roy. Soc.* (A) **140**: 653–660.

93. WOLSKI, P. 1921. Ueber optisch leere Flüssigkeiten. *Kolloidchem. Beih.* **13**: 137–164.

94. YOUNG, S. W. 1911. Mechanical stimulus to crystallization in supercooled liquids. *Jour. Amer. Chem. Soc.* **33**: 148–162.

95. YOUNG, S. W., and W. E. BURKE. 1907. Further studies of supercooled liquids. *Jour. Amer. Chem. Soc.* **29**: 329–339.

96. YOUNG, S. W., and J. P. MITCHELL. 1904. A study of the supercooled fusions and solutions of sodium thiosulphate. *Jour. Amer. Chem. Soc.* **26**: 1389–1413.

97. YOUNG, S. W., and W. J. VAN SICKLEN. 1913. The mechanical stimulus to crystallization. *Jour. Amer. Chem. Soc.* **35**: 1067–1078.

98. ZELDOVICH, J. B. 1943. On the theory of new phase formation; Cavitation. *Acta phys.-chim. U. R. S. S.* **18**: 1–22.

SUPPLEMENTAL REFERENCES

The following references, selected from the very extensive bibliography of this field and not included in the preceding list, may assist future workers in obtaining their initial orientation.

1. BIOLOGICAL OBSERVATIONS

BARNES, T. C., and T. L. JAHN. 1933. The effect of ice and steam water on *Euglena*. *Proc. Natl. Acad. Sci.* **19**: 638–640.

FALEEV, A. V. 1943. The influence of "ice water" and distilled water on the activity of yeast. *Chem. Abstr.* **37**: 4421.

FALEEV, A. V., and F. T. SYKHENKO. 1939. Effect of thawed and distilled water on the activity of blood catalase. *Chem. Abstr.* **33**: 4275.

2. CRYSTAL GROWTH, NATURE OF

BECKER, R., and W. DÖRING. 1935. Kinetische Behandlung der Keimbildung in übersättigten Dämpfen. *Ann. Phys., Lps.* (5) **24**: 719–752.

BIRSTEIN, G., and M. BLUMENTHAL. 1937. Zur Kinetik des Kristallisationsprozesses. *Bull. int. Acad. Cracovie* (A) **1937**: 399–421.

BRANDES, H. 1927. Zur Theorie des Kristallwachstums. *Ztsch. phys. Chem.* **126**: 196–210.

BRANDES, H., and M. VOLMER. 1931. Zur Theorie des Kristallwachstums. *Ztschr. phys. Chem.* (A) **155**: 466–470.

DEHLINGER, U. 1941. Gleichgewicht und Keimbildung beim Schmelzen und Erstarren. *Phys. Ztschr.* **42**: 197–203.

KAISCHEW, R. 1936. Zur Theorie des Kristallwachstums. *Ztschr. Phys.* **102**: 684–690.

KOSSEL, W. 1930. Ueber Krystallwachstum. *Naturwissenschaften* **18**: 901–910.

—— 1943. Gerichtete chemische Vorgänge (Auf- und Abbau von Kristallen). *Angew. Chem.* (n.F.) **56**: 33–41. Now called *Die Chemie*.

ROGINSKY, S. 1939. The kinetics of crystal growth. *Acta phys.-chim. U. R. S. S.* **10**: 825–844.

STRANSKY, I. N. 1928. Zur Theorie des Kristallwachstums. *Ztschr. phys. Chem.* **136**: 259–278.

—— 1942. Ueber das Verhalten nichtpolarer Kristalle dicht unterhalb des Schmelzpunktes und beim Schmelzpunkt selbst. *Ztschr. Phys.* **119**: 22–34.

STRANSKI, I. N., and R. KAISCHEW. 1935. Gleichgewichtsform und Wachstumsform der Kristalle. *Ann. Phys., Lps.* (5) **23**: 330–338.

3. CRYSTAL GROWTH, VELOCITY OF

CAMPBELL, A. N., and A. J. R. CAMPBELL. 1937. The velocity of crystallisation from supersaturated solutions. *Trans. Faraday Soc.* **33**: 299–308.

DREYER, F. 1904. Ueber die Kristallisationsgeschwindigkeit binärer Schmelzen. *Ztschr. phys. Chem.* **48**: 467–482.

FÖRSTER, T. 1936. Ueber die experimentelle Bestimmung der linearen Kristallisationsgeschwindigkeit. *Ztschr. phys. Chem.* (A) **175**: 177–186.

FREUNDLICH, H., and F. OPPENHEIMER. 1925. Ueber die Krystallisationsgeschwindigkeit unterkühlter wässeriger Sole. *Ber. dtsch. chem. Ges.* (B) **58**: 143–148.

FRIEDLÄNDER, J., and G. TAMMANN. 1897. Ueber die Krystallisationsgeschwindigkeit. *Ztschr. phys. Chem.* **24**: 152–159.

GERNEZ, D. 1882. Recherches sur la durée de la solidification des corps surfondus. *C. R. Acad. Sci., Paris* **95**: 1278–1280.

JENKINS, J. D. 1925. The effect of various factors upon the velocity of crystallization of substances from solution. *Jour. Amer. chem. Soc.* **47**: 903–922.

KAISCHEW, R., and I. N. STRANSKI. 1934. Zur Theorie der linearen Kristallisationsgeschwindigkeit. *Ztschr. phys. Chem.* (A) **170**: 295–299.

KÜSTER, F. W. 1898. Ueber die Krystallisationsgeschwindigkeit. *Ztschr. phys. Chem.* **25**: 480–482.

MARC, R. 1908. Ueber die Kristallisation aus wässerigen Lösungen. *Ztschr. phys. Chem.* **61**: 385–398.

—— 1909. Ueber die Kristallisation aus wässerigen Lösungen. Zweite Mitteilung. *Ztschr. phys. Chem.* **67**: 470–500.

NACKEN, R. 1917. Velocity of crystallisation in under-cooled fusions. *Jour. Chem. Soc.* 112,: 363 (abstr.).

PADOA, M., and D. GALEATI. 1905. Sulle diminuzioni della velocità di cristallizzazione provocate da sostanze estranee. *Gazz. chim. ital.* **35**: 181–191.

v.PICKARDT, E. 1902. Die molekulare Verminderung der Krystallisations-Geschwindigkeit durch Zusatz von Fremdstoffen. *Ztschr. phys. Chem.* **42**: 17–49.

TAMMANN, G. 1897. Ueber die Erstarrungsgeschwindigkeit. *Ztschr. phys. Chem.* **23**: 326–328.

TAMMANN, G., and A. BÜCHNER. 1935. Die lineare Kristallisationsgeschwindigkeit des Eises aus gewöhnlichen und schweren Wasser. *Ztschr. anorg. Chem.* **222**: 12–16.

VOLMER, M., and M. MARDER. 1931. Zur Theorie der linearen Kristallisationsgeschwindigkeit unterkühlter Schmelzen und unterkühlter fester Modifikationen. *Ztschr. phys. Chem.* (A) **154**: 97–112.

WALTON, J. H., and R. C. JUDD. 1914. The velocity of the crystallisation of undercooled water. *Jour. phys. Chem.* **18**: 722–728.

WILSON, H. A. 1900. On the velocity of solidification and viscosity of supercooled liquids. *Philos. Mag.* (5) **50**: 238–250.

4. FORMATION OF NUCLEI

AMSLER, J. 1942. Versuche über die Keimbildung in übersättigten Lösungen. *Helv. phys. Acta* **15**: 699–732.

FARKAS, L. 1927. Keimbildungsgeschwindigkeit in übersättigten Dämpfen. *Ztschr. phys. Chem.* **125**: 236–242.

GORSKY, F. K. 1936. Kristallisation von dünnen schichten unterkühlter Flüssigkeiten. *Phys. Ztschr. Sowjet.* **9**: 89–93.

KAISCHEW, R., and I. N. STRANSKI. 1934. Zur kinetischen Ableitung der Keimbildungsgeschwindigkeit. *Ztschr. phys. Chem.* (B) **26**: 317–326.

STAUFF, J. 1940. Keimbildungsgeschwindigkeit von übersättigten Lösungen als Mittel zur Bestimmung von Lösungszuständen. 1. Teil. Lösungen von KClO₃. *Ztschr. phys. Chem.* (A) **187**: 107–118.

STRANSKI, I. N. 1941. Wesen der Keimbildung neuer Phasen. *Verh. dtsch. Ingr. Verfahr.-Tech.* **1941**: 39–43.

STRANSKI, I. N., and D. TOTOMANOW. 1933. Keimbildungsgeschwindigkeit und Ostwaldsche Stufenregel. *Ztschr. phys. Chem.* (A) **163**: 399–408.

5. "MEMORY"

GOETZ, A. 1930. On mechanical and magnetic factors influencing the orientation and perfection of bismuth single-crystals. *Phys. Rev.* (2) **35**: 193–207.

LeBLANC, M., and E. MÖBIUS. 1933. Ist das Schmelzen kristallisierter Körper mit einer Vor- und einer Nachgeschichte verbunden? *Ber. und Ver. sächs. Akad. Wiss.* (Math.-phys. Kl.) **85**: 75–96.

ROGINSKY, S., L. SENA, and J. ZELDOWITSCH. 1932. Beitrag zum Mechanismus der Erscheinung des "Gedächtnisses" der widerholten Kristallisation. *Phys. Ztschr. Sowjet.* **1**: 630–639.

6. PERSISTENCE OF CRYSTALS IN MELT

BOYDSTON, R. W. 1927. Thermo-electric effect in single-crystal bismuth. *Phys. Rev.* (2) **30**: 911–921.

BURK, R. E. 1935. The significance of the persistence of the crystalline state above the melting point. *Science* (n.s.) **81**: 344–345.

DONAT, E., and A. STIERSTADT. 1933. Ueber flüssige Metalleinkristalle. I. *Ann. Phys., Lpz.* (5) **17**: 897–914.

SOROOS, A. 1932. Thermoelectric power of single crystal bismuth near the melting point. *Phys. Rev.* (2) **41**: 516–522.

7. SIZE OF MOTES, ETC.

DANKOV, N., and A. KOČETKOV. 1934. On limiting dimensions of catalyst particles. *C. R. Acad. Sci. U. R. S. S.* **2**: 362–364.

HAMBURGER, I. L. 1938. Oververzadigde oplossingen, in het bijzonder van calciumzouten. *Chem. Weekbl.* **35**: 886–906.

JONES, M., and J. R. PARTINGTON. 1915. Experiments on supersaturated solutions. *Jour. Chem. Soc.* **107**: 1019–1025.

JONES, W. J., and J. R. PARTINGTON. 1915. A theory of supersaturation. *Philos. Mag.* (6) **29**: 35–40.

KÜSTER, F. W. 1903. Ueber das Wesen des metastabilen Zustandes. *Ztschr. anorg. Chem.* **33**: 363–368.

RAO, S. R. 1931. Diamagnetism and the colloidal state. *Indian Jour. Phys.* **6**: 241–259.

RIE, E. 1921. Oberflächenspannung und Aggregatzustandsänderungen. *Verh. dtsch. phys. Ges.* (3) **2**: 33–34.

8. STABILITY

ANDRONIKASCHWILI, E. 1937. Ueber die Bildung von Zentren der Phasenumwandlungen in kondensierten Systemen. *Acta phys.-chim. U. R. S. S.* **6**: 689–700.

DEHLINGER, U. 1941. Gleichgewicht und Keimbildung beim Schmelzen und Erstarren. *Phys. Ztschr.* **42**: 197–203.

9. X-RAY DATA

BRESLER, S. 1939. The molecular-statistic theory of melting. *Acta phys.-chim. U. R. S. S.* **10**: 491–512.

KATZOFF, S. 1934. X-ray studies of the molecular arrangement in liquids. *Jour. Chem. Phys.* **2**: 841–851.

LENNARD-JONES, J. E., and A. F. DEVONSHIRE. 1939. Critical and coöperative phenomena. III. A theory of melting and the structure of liquids. *Proc. Roy. Soc.* **169**: 317–338.

MORGAN, J., and B. E. WARREN. 1938. X-ray analysis of the structure of water. *Jour. Chem. Phys.* **6**: 666–673.

PENNYCUICK, S. W. 1928. The structure of water. *Jour. Phys. Chem.* **32**: 1681–1696.

PRIETZSCHK, A. 1941. Röntgenuntersuchungen an unterkühlten Aethylalkohol. *Ztschr. Phys.* **117**: 482–501.

STEWART, G. W. 1930. The cybotactic (molecular group) condition in liquids; the nature of the association of octyl alcohol molecules. *Phys. Rev.* (2) **35**: 726–732.

——— 1930. X-ray diffraction in water 2° to 98°C: The nature of molecular association. *Phys. Rev.* (2) **35**: 1426 (abstr.).

——— 1931. X-ray diffraction in water: the nature of molecular association. *Phys. Rev.* (2) **37**: 9–16.

——— 1932. Evidence for the cybotactic group view of the interior of a liquid. *Indian Jour. Phys.* **7**: 603–615.

——— 1941. Effect of ions on the liquid structure of water. *Nature* **148**: 698 (Research item).

INDEX